Simplified Mechanics
and Strength of Materials

Simplified Mechanics
and Strength of Materials

||||||||||||||||||||||||||||||||||||||||||||||||||||||||||||||||||

The Late Harry Parker, M.S.
Formerly Professor of Architectural Construction
University of Pennsylvania

FOURTH EDITION

Prepared By

JAMES AMBROSE, M.S.
Professor of Architecture
University of Southern California

A Wiley-Interscience Publication
JOHN WILEY & SONS
New York · Chichester · Brisbane · Toronto · Singapore

Library of Congress Cataloging in Publication Data:

Parker, Harry, 1887-
  Simplified mechanics and strength of materials.

  "A Wiley-Interscience publication."
  Includes index.
  1. Mechanics, Applied. 2. Strength of materials.
I. Ambrose, James E. II. Title.

TA350.P3 1986   620.1   86-11004
ISBN 0-471-82269-8

Printed in the United States of America

10 9 8 7 6 5 4 3 2 1

Preface

||||||||||||||||||||||||||||||||||||||||||||||||||

The purpose and scope of this book were stated by Professor Parker in the preface to the first edition, which is reprinted on the following pages.

The arguments for a "simplified" treatment of the basic topics of structural mechanics and strength of materials are as valid today as when Professor Parker made them in 1951. Studies and work in this field have grown increasingly sophisticated, yet the basic principles and the work required for investigation and design of most common elements of ordinary building structures are limited and elementary. This new edition retains the spirit and basic style of the original work, while offering a somewhat updated and slightly broadened treatment. Several topics have been added, including sliding friction, three-dimensional force systems, combined stresses, and rigid frames. For most of the work the mathematical treatment is limited to the use of algebra and geometry. In a few instances, however, elementary trigonometry has been used where it offers a more practical procedure.

In Professor Parker's early days of teaching both architecture and engineering students were extensively trained in drafting. The use of graphic techniques for force vector analysis was therefore popular. Sadly, the teaching of drafting has considerably declined, so that its use for analysis work is less practical. For most of the applied work in later chapters the computations are done entirely algebraically, although the development of graphic techniques has been retained in the early chapters.

I am grateful to the American Institute of Steel Construction, the National Forest Products Association, and the American Concrete Institute for their permission to reproduce and adapt materials from their publications. I am also grateful to the editors and production staff at

John Wiley and Sons for their patient and thorough work and their maintenance of high standards through the editing and production processes.

As usual, I am in debt to the many students who have passed (or not) through the courses that I have taught over the last 25 years; I have benefited selfishly from both their successes and their failures. Finally, I am grateful to my colleagues, my present students, and my family for their endurance and tolerance in the face of my absorption with this work.

JAMES AMBROSE

Westlake Village, California
June 1986

Preface to the First Edition
IIIIIIIIIIIIIIIIIIIIIIIIIIIIIIIIIIIIIIIIIIIIIIIIIIIIIIIII

Since engineering design is based on the science of mechanics, it is impossible to overemphasize the importance of a thorough knowledge of this basic subject. Regardless of the particular field of engineering in which a student is interested, it is essential that he understand fully the fundamental principles that deal with the action of forces on bodies and the resulting stresses.

Each of the other volumes of this "simplified" series has, in general, dealt with the design of a particular material, structural steel, timber, and reinforced concrete. In each of these books is included only pertinent material, relating to the principles of mechanics, in the accompanying discussions and explanations. Obviously, such discussions and explanations are brief and limited in scope, and many important items are necessarily omitted. Students in engineering will find that the knowledge gained from a study of mechanics and strength of materials will enable them to understand more fully the theory of the design of structural members regardless of the material involved.

This is an elementary treatment written for those who have had limited preparation. The best books on the subject of mechanics and strength of materials make use of physics, calculus, and trigonometry. Such books are useless for many ambitious men. Consequently, this book has been prepared for the student who has not obtained a practical appreciation of mechanics or advanced mathematics. A working knowledge of algebra and arithmetic is sufficient to enable him to comprehend the mathematics involved in this volume.

This book has been written for use as a textbook in courses in mechanics and strength of materials and for use by practical men interested in mechanics and construction. Because it is elementary, the material has been arranged so that it may be used for home study. For those who have had previous training it will serve as a refresher course in reviewing the most important of the basic principles of structural design.

One of the most important features of this book is a detailed explanation of numerous illustrative examples. In so far as possible, the examples relate to problems encountered in practice. The explanations are followed by problems to be solved by the student.

The designer of structural members must have at hand tables of allowable stresses, properties of sections, and other tables giving engineering data. Such tables are included in this book, and reference books are not required. The author is indebted to The American Concrete Institute, The American Institute of Steel Construction, The National Lumber Manufacturers Association, and the Timber Engineering Company for their kindness and cooperation in granting permission to reproduce tables and other data from their publications.

This book presents no short-cuts to a knowledge of the fundamental principles of mechanics and strength of materials. There is nothing unique in the presentation, for the discussions follow accepted present-day design procedure. It is the belief of the author, however, that a thorough understanding of the material contained herein will afford a foundation of practical information and serve as a step to further study.

HARRY PARKER

High Hollow
Southampton
Bucks County, Pennsylvania
May 1951

Contents

||||||||||||||||||||||||||||||||||||||||||||

Simplified Mechanics
and Strength of Materials

1

Introduction

||||||||||||||||||||||||||||||||||||||||||||||||||

1-1. Mechanics

The science of *mechanics* concerns the actions of forces on material bodies. Most of engineering design is based on applications of the science of mechanics. *Statics* is the branch of mechanics that deals with forces in equilibrium or with bodies held motionless by the forces acting on them. *Dynamics* is the branch of mechanics that concerns bodies in motion or forces that are involved with time-dependent relationships.

1-2. Strength of Materials

When forces act on a material body, two things happen. First, internal forces that resist the effects of the external forces are set up in the body. These internal forces produce *stresses* in the material of the body. Second, the external forces produce *deformations* or changes in the shape of the body.

Strength of materials, or mechanics of materials, is the study of the properties of material bodies that enable them to resist the actions of external forces, of the stresses within the bodies, and of the deformations that result from the external forces.

1-3. Structural Mechanics

In architectural and civil engineering the related subjects of statics and strength of materials are often given the overall designation *structural mechanics* since they form the basis of structural design.

In general, an architect or engineer is confronted with two distinct types of problems, *design* and *investigation*. Design problems are problems in which the material, shape, and size of a body are to be determined in order that external forces may be resisted economically. Problems of investigation give as data the kind of material and its size and shape as well as the loads to be resisted by the body. The architect or engineer computes the magnitudes of the internal resisting forces (stresses) set up in the body in order to determine whether or not the size of the member is sufficiently large.

1-4. Units of Measurement

At the time of preparation of this edition, the building industry in the United States is still in a state of confused transition from the use of English units (feet, pounds, etc.) to the new metric-based system referred to as the SI units (for Système International). Although a complete phase-over to SI units seems inevitable, at the time of this writing the construction-materials and products suppliers in the United States are still resisting it. Consequently, the AISC Manual and most building codes and other widely used references are still in the old units. (The old system is now more appropriately called the U.S. system because England no longer uses it!) Although it results in some degree of clumsiness in the work, we have chosen to give the data and computations in this book in both units as much as is practicable. The technique is generally to perform the work in U.S. units and immediately follow it with the equivalent work in SI units enclosed in brackets [thus] for separation and identity.

Table 1-1 lists the standard units of measurement in the U.S. system with the abbreviations used in this work and a description of the type of the use in structural work. In similar form Table 1-2 gives the corresponding units in the SI system. The conversion units used in shifting from one system to the other are given in Table 1-3.

For some of the work in this book, the units of measurement are not significant. What is required in such cases is simply to find a nu-

TABLE 1-1. Units of Measurement: U.S. System

Name of Unit	Abbreviation	Use
Length		
Foot	ft	large dimensions, building plans, beam spans
Inch	in.	small dimensions, size of member cross sections
Area		
Square feet	ft^2	large areas
Square inches	$in.^2$	small areas, properties of cross sections
Volume		
Cubic feet	ft^3	large volumes, quantities of materials
Cubic inches	$in.^3$	small volumes
Force, mass		
Pound	lb	specific weight, force, load
Kip	k	1000 pounds
Pounds per foot	lb/ft	linear load (as on a beam)
Kips per foot	k/ft	linear load (as on a beam)
Pounds per square foot	lb/ft^2, psf	distributed load on a surface
Kips per square foot	k/ft^2, ksf	distributed load on a surface
Pounds per cubic foot	lb/ft^3, pcf	relative density, weight
Moment		
Foot-pounds	ft-lb	rotational or bending moment
Inch-pounds	in.-lb	rotational or bending moment
Kip-feet	k-ft	rotational or bending moment
Kip-inches	k-in.	rotational or bending moment
Stress		
Pounds per square foot	lb/ft^2, psf	soil pressure
Pounds per square inch	$lb/in.^2$, psi	stresses in structures
Kips per square foot	k/ft^2, ksf	soil pressure
Kips per square inch	$k/in.^2$, ksi	stresses in structures
Temperature		
Degree Fahrenheit	°F	temperature

TABLE 1-2. Units of Measurement: SI System

Name of Unit	Abbreviation	Use
Length		
Meter	m	large dimensions, building plans, beam spans
Millimeter	mm	small dimensions, size of member cross sections
Area		
Square meters	m^2	large areas
Square millimeters	mm^2	small areas, properties of cross sections
Volume		
Cubic meters	m^3	large volumes
Cubic millimeters	mm^3	small volumes
Mass		
Kilogram	kg	mass of materials (equivalent to weight in U.S. system)
Kilograms per cubic meter	kg/m^3	density
Force (load on structures)		
Newton	N	force or load
Kilonewton	kN	1000 newtons
Stress		
Pascal	Pa	stress or pressure (1 pascal = $1 N/m^2$)
Kilopascal	kPa	1000 pascals
Megapascal	MPa	1,000,000 pascals
Gigapascal	GPa	1,000,000,000 pascals
Temperature		
Degree Celsius	°C	temperature

merical answer. The visualization of the problem, the manipulation of the mathematical processes for the solution, and the quantification of the answer are not related to the specific units—only to their relative values. In such situations we have occasionally chosen not to present the work in dual units, to provide a less confusing illustration for the reader. Although this procedure may be allowed for the learning ex-

TABLE 1-3.   Factors for Conversion of Units

To Convert from U.S. Units to SI Units Multiply by	U.S. Unit	SI Unit	To Convert from SI Units to U.S. Units Multiply by
25.4	in.	mm	0.03937
0.3048	ft	m	3.281
645.2	in.^{2}	mm^2	1.550×10^{-3}
16.39×10^3	in.^{3}	mm^3	61.02×10^{-6}
416.2×10^3	in.^{4}	mm^4	2.403×10^{-6}
0.09290	ft^2	m^2	10.76
0.02832	ft^3	m^3	35.31
0.4536	lb (mass)	kg	2.205
4.448	lb (force)	N	0.2248
4.448	kip (force)	kN	0.2248
1.356	ft-lb (moment)	N-m	0.7376
1.356	kip-ft (moment)	kN-m	0.7376
1.488	lb/ft (mass)	kg/m	0.6720
14.59	lb/ft (load)	N/m	0.06853
14.59	kips/ft (load)	kN/m	0.06853
6.895	psi (stress)	kPa	0.1450
6.895	ksi (stress)	MPa	0.1450
0.04788	psf (load or pressure)	kPa	20.93
47.88	ksf (load or pressure)	kPa	0.02093
$0.566 \times (°F - 32)$	°F	°C	$(1.8 \times °C) + 32$

ercises in this book, the structural designer is generally advised to develop the habit of always indicating the units for any numerical answers in structural computations.

1-5.   Computations

In most professional design firms structural computations are most commonly done with computers, particularly when the work is complex or repetitive. Anyone aspiring to participation in professional design work is advised to acquire the background and experience necessary to the application of computer-aided techniques. The computational work in this book is simple and can be performed easily with a pocket calculator. The reader who has not already done so is advised to obtain one. The "scientific" type with eight-digit capacity is quite sufficient.

For the most part, structural computations can be rounded off. Accuracy beyond the third place is seldom significant, and this is the level used in this work. In some examples more accuracy is carried in early stages of the computation to ensure the desired degree in the final answer. All the work in this book, however, was performed on an eight-digit pocket calculator.

1-6. Symbols

The following "shorthand" symbols are frequently used.

Symbol	Reading
$>$	is greater than
$<$	is less than
$\geqq$	equal to or greater than
$\leqq$	equal to or less than
$6'$	6 feet
$6''$	6 inches
Σ	the sum of
ΔL	change in L

1-7. Notation

Use of standard notation in the general development of work in mechanics and strength of materials is complicated by the fact that there is some lack of consistency in the notation currently used in the field of structural design. Some of the standards used in the field are developed by individual groups (notably those relating to a single basic material, wood, steel, concrete, masonry, etc.) which each have their own particular notation. Thus the same type of stress (such as shear stress in a beam) or the same symbol (f_c) may have various representations in structural computations. To keep some form of consistency in this book we use the following notation, most of which is in general agreement with that used in structural design work at present.

a (1) Moment arm; (2) acceleration; (3) increment of an area

A Gross (total) area of a surface or a cross section

b Width of a beam cross section

B Bending coefficient

c Distance from neutral axis to edge of a beam cross section

d Depth of a beam cross section or overall depth (height) of a truss

D (1) Diameter; (2) deflection

e (1) Eccentricity (dimension of the mislocation of a load resultant from the neutral axis, centroid, or simple center of the loaded object); (2) elongation

E Modulus of elasticity (ratio of unit stress to the accompanying unit strain)

f Computed unit stress

F (1) Force; (2) allowable unit stress

g Acceleration due to gravity

G Shear modulus of elasticity

h Height

H Horizontal component of a force

I Moment of inertia (second moment of an area about an axis in the plane of the area)

J Torsional (polar) moment of inertia

K Effective length factor for slenderness (of a column: KL/r)

M Moment

n Modular ratio (of the moduli of elasticity of two different materials)

N Number of

p (1) Percent; (2) unit pressure

P Concentrated load (force at a point)

r Radius of gyration of a cross section

R Radius (of a circle, etc.)

s (1) Center-to-center spacing of a set of objects; (2) distance of travel (displacement) of a moving object; (3) strain or unit deformation

t (1) Thickness; (2) time

T (1) Temperature; (2) torsional moment

V (1) Gross (total) shear force; (2) vertical component of a force

w (1) Width; (2) unit of a uniformly distributed load on a beam

W (1) Gross (total) value of a uniformly distributed load on a beam;
(2) gross (total) weight of an object

Δ (delta)	Change of
Σ (sigma)	Sum of
θ (theta)	Angle
μ (mu)	Coefficient of friction
ϕ (phi)	Angle

2

Forces

||||||||||||||||||||||||||||||||||||||||||||

2-1. General

The idea of force is one of the fundamental concepts of mechanics and as such does not lend itself to simple, precise definition. For our purposes at this stage of study, we may define a force as that which produces, or tends to produce, motion or a change in the motion of bodies. One type of force is the effect of *gravity*, by which all bodies are attracted toward the center of the earth. The magnitude of the force of gravity is the *weight* of a body. The amount of material in a body is its *mass*. In the U.S. (old English) System the force effect of gravity is equated to the weight. In SI units a distinction is made between weight and force which results in the force unit of a *newton*. In the U.S. System the basic unit of force is the *pound*, although in engineering work a commonly used unit is the *kip* (1000 pounds or, literally, a kilopound).

Figure 2-1*a* represents a block of metal weighing 6400 1b supported on a short piece of wood having an 8 X 8 in. cross-sectional area. The wood is in turn supported on a base of masonry. The force of the metal block exerted on the wood is 6400 lb, or 6.4 kips. Note that the wood transfers a force of equal magnitude (ignoring the weight of the wood block) to the masonry base. If there is no motion (equilibrium), there must be an equal upward force in the base. For equilibrium, force actions must exist in opposed pairs. In this instance the magnitude of

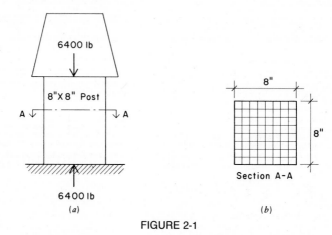

FIGURE 2-1

the force is 6400 lb, and the resisting force offered by the masonry and the wood block is also 6400 lb. The resisting force in the block of wood is developed by *stress*, defined as internal force per unit area of the block's cross section. For the situation shown, each square inch of the cross section must develop a stress equal to $6400/64 = 100$ lb/in.[2] (psi). (see Figure 2-1*b*.)

2-2. Types of Force

External forces may result from a number of sources but essentially are distinguished only as being either static or dynamic. At present we are dealing only with static forces. Internal forces are one of three possible types, tension, compression, or shear.

When a force acts on a body in a manner that tends to shorten the body or to push the parts of the body together, the force is a compressive force and the stresses within the body are compressive stresses. The block of metal acting on the piece of wood in Figure 2-1 represents a compressive force, and the resulting stresses in the wood are compressive stresses.

Figure 2-2 represents a 0.5-in.-diameter steel rod suspended from a ceiling. A weight of 1500 lb is attached to the lower end of the rod. The weight constitutes a tensile force, which is a force that tends to lengthen or pull apart the body on which it acts. In this example the

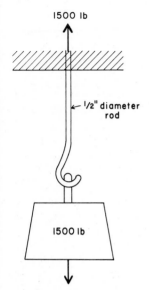

1500 lb

½" diameter rod

1500 lb

FIGURE 2-2

rod has a cross-sectional area of πR^2, or $3.1416(0.25)^2 = 0.196$ in.^{2}. Hence the tensile unit stress in the rod is $1500/0.196 = 7653$ psi.

In this book we consider the weights given in U.S. units to be forces and make direct conversion from pounds of force to newtons of force. (See the discussion in Chapter 1 and the conversion factors given in Table 1-3.) Thus for the wood block in Figure 2-1:

Force = 6400 lb = 4.448 × 6400 = 28,467 N,   or 28.467 kN

Stress = 100 psi = 6.895 × 100 = 689.5 kPa

Consider the two steel bars held together by a 0.75-in.-diameter bolt as shown in Figure 2-3. The force exerted on the bolt is 5000 lb. In addition to the tension in the bars and the bearing action of the bars on the bolt, there is a tendency for the bolt to fail by a cutting action at the plane at which the two bars are in contact. This force action is called *shear*; it results when two parallel forces having opposite sense of direction act on a body, tending to cause one part of the body to slide past an adjacent part. The bolt has a cross-sectional area of $3.1416(0.75)^2/4 = 0.4418$ in.^{2} [285 mm^2] , and the unit shear stress is equal to $5000/0.4418 = 11,317$ psi [78.03 MPa]. Note particularly

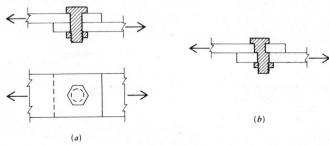

FIGURE 2-3

that this example illustrates the computation of shear stress and that the magnitude of this stress would be the same if the forces on the bars were reversed in sense, producing compression instead of tension in the bars.

2-3. Vectors

A quantity that combines both magnitude and direction is a *vector* quantity, whereas a *scalar* quantity involves magnitude but no direction. Forces, velocity, and acceleration are vector quantities, and energy, time, and temperature are scalar quantities. A vector may be represented graphically by a drawn line; the solutions of problems involving forces may thus sometimes be accomplished by the construction of diagrams, with the forces being represented by straight lines.

2-4. Properties of Forces

To identify a force it is necessary to establish the following:

Magnitude, or the amount, of the force, which is measured in weight units such as pounds or tons.

Direction of the force, which refers to the orientation of its path or line of action. Direction is usually described by the angle that the line of action makes with some reference, such as the horizontal.

Sense of the force, which refers to the manner in which it acts along its line of action (e.g., up or down). Sense is usually expressed algebraically in terms of the sign of the force, either plus or minus.

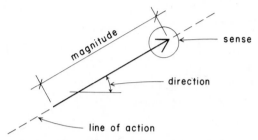

FIGURE 2-4.   Graphical representation of a force.

Forces can be represented graphically in terms of these three properties by the use of an arrow, as shown in Figure 2-4. Drawn to some scale, the length of the arrow represents the magnitude of the force. The angle of inclination of the arrow represents the direction of the force. The location of the arrowhead determines the sense of the force. This form of representation can be more than merely symbolic, since actual mathematical manipulations may be performed using the vector representation that the force arrows constitute. In the work in this book arrows are used in a symbolic way for visual reference when performing algebraic computations, and in a truly representative way when performing graphical analyses.

In addition to the basic properties of magnitude, direction, and sense, some other concerns that may be significant for certain investigations are

The *position of the line of action* of the force with respect to the lines of action of other forces or to some object on which the force operates, as shown in Figure 2-5.

The *point of application* of the force along its line of action may be of concern in analyzing for the specific effect of the force on an object, as shown in Figure 2-6.

When forces are not resisted, they tend to produce motion. An inherent aspect of static forces is that they exist in a state of *static equilibrium*, that is, with no motion occurring. In order for static equilibrium to exist, it is necessary to have a balanced system of forces. An important consideration in the analysis of static forces is the nature of the geo-

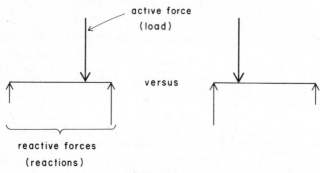

active force
(load)

versus

reactive forces
(reactions)

FIGURE 2-5

metric arrangement of the forces in a given set of forces that constitute a single system. The usual technique for classifying force systems involves consideration of whether the forces in the system are

Coplanar.   All acting in a single plane, such as the plane of a vertical wall.
Parallel.   All having the same direction.
Concurrent.   All having their lines of action intersect at a common point.

Using these three considerations, the possible variations are given in Table 2.1 and illustrated in Figure 2-7. Note that variation 5 in the table is really not possible, since a set of coacting forces that is parallel and concurrent cannot be noncoplanar; in fact, they all fall on a single line of action and are called collinear.

It is necessary to qualify a set of forces in the manner just illustrated before proceeding with any analysis, whether it is to be performed algebraically or graphically.

versus

FIGURE 2-6.   Effect of point of application of a force.

TABLE 2-1.   Classification of Force Systems[a]

System Variation	Qualifications		
	Coplanar	Parallel	Concurrent
1	yes	yes	yes
2	yes	yes	no
3	yes	no	yes
4	yes	no	no
5	no[b]	yes	yes
6	no	yes	no
7	no	no	yes
8	no	no	no

[a] See Figure 2-7.
[b] Not possible if forces are parallel and concurrent.

2-5.   Motion

A force was defined earlier as that which produces or tends to produce motion or a change of motion of bodies. *Motion* is a change of position with respect to some object regarded as having a fixed position. When the path of a moving point is a straight line, the point has *motion of translation*. When the path of a point is curved, the point has *curvilin-*

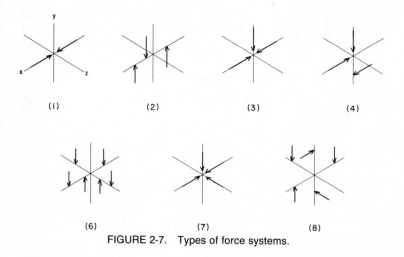

FIGURE 2-7.   Types of force systems.

ear motion or *motion of rotation*. When the path of a point lies in a plane, the point has *plane motion*. Other motions are *space motions*.

2-6. Static Equilibrium

A body is in *equilibrium* when it is either at rest or has uniform motion. When a system of forces acting on a body produces no motion in the body, the system of forces is said to be in equilibrium.

A simple example of equilibrium is illustrated in Figure 2-8a. Two equal, opposite, and parallel forces, having the same line of action, P_1 and P_2, act on a body. We say that the two forces *balance* each other; the body does not move and the system of forces is in equilibrium. These two forces are *concurrent*. If the lines of action of a system of forces have a point in common, the forces are concurrent.

Another example of forces in equilibrium is illustrated in Figure 2-8b. A vertical downward force of 300 lb acts at the mid-point in the length of a beam. The two upward vertical forces of 150 lb each (the reactions) act at the ends of the beam. The system of three forces is in equilibrium. The forces are parallel and, not having a point in common, *nonconcurrent*.

2-7. Resultant of Forces

The *resultant* of a system of forces is the simplest system (usually a single force) that has the same effect as the various forces in the system acting simultaneously. The lines of action of any system of two non-parallel forces must have a point in common, and the resultant of the two forces will pass through this common point. The resultant of two

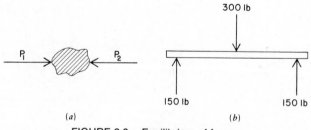

(a)   (b)

FIGURE 2-8.   Equilibrium of forces.

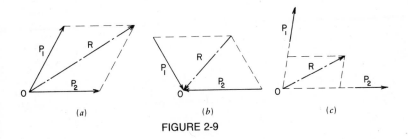

FIGURE 2-9

nonparallel forces may be found graphically by constructing a *parallelogram of forces*.

This graphical construction is based on the *parallelogram law* which may be stated thus: two nonparallel forces are laid off at any scale (of so many pounds to the inch) with both forces pointing toward or both forces pointing away from the point of intersection of their lines of action. A parallelogram is then constructed with the two forces as adjacent sides. The diagonal of the parallelogram passing through the common point is the resultant in magnitude, direction, and line of action, the direction of the resultant being similar to that of the given forces, toward or away from the point in common. In Figure 2-9a P_1 and P_2 represent two nonparallel forces whose lines of action intersect at point O. The parallelogram is drawn, and the diagonal R is the resultant of the given system. In this illustration note that the two forces point *away* from the point in common, hence the resultant also has its direction away from point O. It is a force upward to the right. Notice that the resultant of forces P_1 and P_2 shown in Figure 2-9b is R; its direction is *toward* the point in common.

Forces may be considered to act at any points on their lines of action. In Figure 2-9c the lines of action of the two forces P_1 and P_2 are extended until they meet at point O. At this point the parallelogram of forces is constructed, and R, the diagonal, is the resultant of the forces P_1 and P_2. In determining the magnitude of the resultant, the scale used is, of course, the same scale used in laying off the given system of forces.

Example 1. A vertical forces of 50 lb and a horizontal force of 100 lb, as shown in Figure 2-10a, have an angle of 90° between their lines of action. Determine the resultant.
Solution: The two forces are laid off from their point of intersection

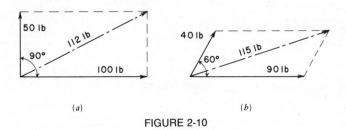

FIGURE 2-10

at a scale of 1 in. = 80 lb. The parallelogram is drawn, and the diagonal is the resultant. Its magnitude scales approximately 112 lb, its direction is upward to the right, and its line of action passes through the point of intersection of the lines of action of the two given forces. By use of a protractor it is found that the angle between the resultant and the force of 100 lb is approximately 26.5°.

Example 2.   The angle between two forces of 40 and 90 lb, as shown in Figure 2-10*b*, is 60°. Determine the resultant.
Solution: The forces are laid off from their point of intersection at a scale of 1 in. = 80 lb. The parallelogram of forces is constructed, and the resultant is found to be a force of approximately 115 lb, its direction is upward to the right, and its line of action passes through the

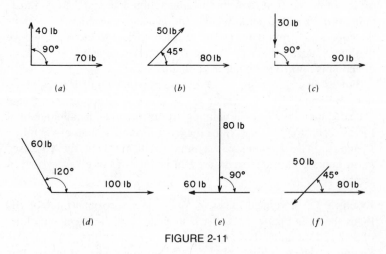

FIGURE 2-11

common point of the two given forces. The angle between the resultant and the force of 90 lb is approximately 17.5°.

Attention is called to the fact that these two problems have been solved graphically by the construction of diagrams. Mathematics might have been employed. For many practical problems graphical solutions give sufficiently accurate answers and frequently require far less time. Do not make diagrams too small. Remember that greater accuracy is obtained by using larger parallograms of forces.

Problems 2-7-A-B-C*-D-E*-F.   By constructing the parallelogram of forces, determine the resultants for the pairs of forces shown in Figures 2-11a, b, c, d, e, and f. Answers to problems followed by an asterisk (*) are given at the end of the book.

2-8. Components of a Force

In addition to combining forces to obtain their resultant, it is often necessary to replace a single force by its *components*. Thus the components of a force are the two or more forces that, acting together, have the same effect as the given force. In Figure 2-10a, if we are *given* the force of 112 lb, its vertical component is 50 lb and its horizontal component is 100 lb. That is, the 112-lb force has been *resolved* into its vertical and horizontal components. Any force may be considered as the resultant of its components.

2-9. Combined Resultants

The resultant of more than two nonparallel forces may be obtained by finding the resultants of pairs of forces and finally the resultant of the resultants.

Example.   Let it be required to find the resultant of the concurrent forces P_1, P_2, P_3, and P_4 shown in Figure 2-12.
Solution: By constructing a parallelogram of forces, the resultant of P_1 and P_2 is found to be R_1. Similarly, the resultant of P_3 and P_4 is R_2. Finally, the resultant of R_1 and R_2 is R, the resultant of the four given forces.

Problems 2-9-A*-B-C.   Using graphical methods, find the resultants of the systems of concurrent forces shown in Figure 2-13a, b, and c.

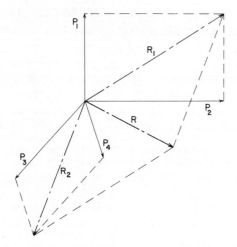

FIGURE 2-12

2-10. Equilibrant

The force required to maintain a system of forces in equilibrium is called the *equilibrant* of the system. Suppose that we are required to investigate the system of two forces, P_1 and P_2, as shown in Figure 2-14. The parallelogram of forces is constructed, and the resultant is found to be R. The system is not in equilibrium. The force required to maintain equilibrium is force E, shown by the dotted line. E, the equilibrant, is equal to the resultant in magnitude and opposite in direction and has the same line of action. The three forces, P_1 and P_2, and E, constitute a system in equilibrium.

If two forces are in equilibrium, they must be equal in magnitude

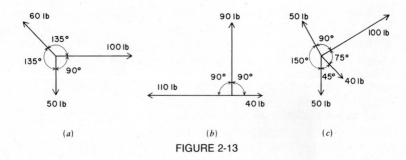

FIGURE 2-13

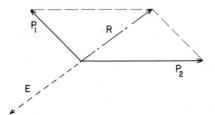

FIGURE 2-14

and opposite in direction and have the same line of action. Either of the two forces may be said to be the equilibrant of the other. The resultant of a system of forces in equilibrium is zero.

2-11. Force Polygon

The resultant of a system of concurrent forces may be found by constructing a *force polygon*. To draw the force polygon, begin with a point and lay off, at a convenient scale, a line parallel to one of the forces, equal to it in magnitude, and having the same direction. From the termination of this line draw similarly another line corresponding to one of the remaining forces and continue in the same manner until all the forces in the given system are accounted for. If the polygon does not close, the system of forces is not in equilibrium, and the line required to close the polygon *drawn from the starting point* is the resultant in magnitude and direction. If the forces in the given system are concurrent, the line of action of the resultant passes through the point they have in common. If they are not concurrent, the line of action of the resultant may be found by constructing a funicular polygon as explained in Section 2-15.

If the force polygon for a system of concurrent forces closes, the system is in equilibrium and the resultant is zero.

Example. Let it be required to find the resultant of the four concurrent forces P_1, P_2, P_3, and P_4 shown in Figure 2-15a. This diagram is called the *space diagram*; it shows the relative positions of the forces in a given system.

Solution: Beginning with some point such as O, shown in Figure 2-15b, draw the upward force P_1. At the upper extremity of the line representing P_1 draw P_2, continuing in a like manner with P_3 and P_4.

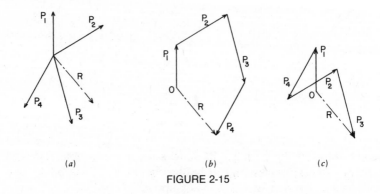

FIGURE 2-15

The polygon does not close; therefore the system is not in equilibrium. The resultant R, shown by the dot-and-dash line, is the resultant of the given system. Note that its direction is *from* the starting point O, downward to the right. The line of action of the resultant of the given system shown in Figure 2-15a has its line of action passing through the point they have in common, its magnitude and direction having been found in the force polygon.

In drawing the force polygon, the forces may be taken in any sequence. In Figure 2-15c a different sequence is taken, but the resultant R is found to have the same magnitude and direction as previously found in Figure 2-15b.

2-12. Bow's Notation

Thus far forces have been identified by the symbols P_1, P_2, and so on. A system of identifying forces, known as Bow's notation, affords many advantages. In this system letters are placed in the space diagram on each side of a force and a force is identified by two letters. *The sequence in which the letters are read is important.* Figure 2-16a shows the space diagram of five concurrent forces. Reading about the point in common *in a clockwise manner* the forces are AB, BC, CD, DE, and EA. When a force in the force polygon is represented by a line, a letter is placed at each end of the line. As an example, the vertical upward force in Figure 2-16a is read AB (note that this is read clockwise about the common point); in the force polygon (Figure 2-16b) the letter a is placed at the bottom of the line representing the force AB

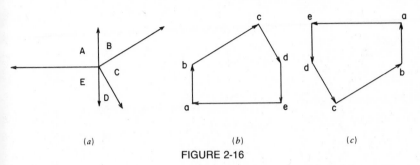

and the letter *b* is at the top. We use capital letters to identify the forces in the space diagrams and lowercase letters in the force polygon. From point *b* in the force polygon we draw force *bc*, then *cd*, and continue with *de* and *ea*. Since the force polygon closes, the five concurrent forces are in equilibrium.

In reading forces, a clockwise manner is used in all the following discussions. It is important that this method of identifying forces be thoroughly understood. To make this clear, suppose that we are asked to draw the force polygon for the five forces shown in Figure 16*a*, *reading the forces counterclockwise*. The the vertical upward force is read *BA* (not *AB* as before), and the force polygon is shown in Figure 2-16*c*. Either direction may be employed, but, to avoid confusion we read forces in a clockwise fashion.

2-13.   Use of the Force Polygon

Two ropes are attached to a ceiling and their ends connected to a ring, making the angles shown in Figure 2-17*a*. A weight of 100 lb is suspended from the ring. Obviously, the stress in the rope *AB* is 100 lb, but the magnitudes of the stresses in ropes *BC* and *CA* are unknown.

The stresses in the ropes *AB*, *BC*, and *CA* constitute three concurrent forces in equilibrium. The magnitude of only one of the forces is known; it is 100 lb in rope *AB*. Since the three concurrent forces are in equilibrium, we know that their force polygon must close, and this fact enables us to find their magnitudes. Let us construct the force polygon. At a convenient scale, draw the line *ab* (Figure 2-17*c*) representing the downward force *AB*, 100 lb. The line *ab* is one side of the force polygon. From point *b* draw a line parallel to rope *BC*; point

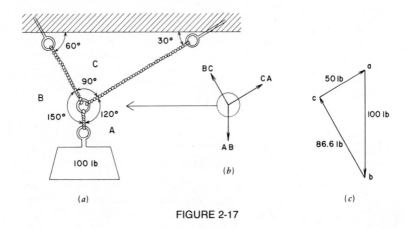

FIGURE 2-17

c will be at some location on this line. Next, draw a line through point a parallel to rope CA; point c will be at some position on this line. Since point c is also on the line through b parallel to BC, the intersection of the two lines determines point c. The force polygon for the three forces is now completed; it is abc, and the lengths of the sides of the polygon represent the magnitudes of the stresses in ropes BC and CA, 86.6 and 50 lb, respectively.

Particular attention is called to the fact that the lengths of the ropes in Figure 2-17a are not an indication of magnitude of the stresses within the ropes; the magnitudes are determined by the lengths of the corresponding sides of the force polygon (Figure 2-17c).

2-14. Stresses in Frames

Figure 2-18a shows a hinged triangular frame with sides AO, BO, and CO attached to three ropes. The ropes have the same angles with the horizontal as corresponding ones in Figure 2-17a. The 100-lb weight is again suspended by the rope AB, and forces are now transferred to ropes BC and CA by the members of the frame.

The stresses in ropes AB, BC, and CA are external forces with respect to the frame, and, since they are in equilibrium, their force polygon must close. This polygon is the triangle abc in Figure 2-18b and it is exactly the same as the one shown in Figure 2-17c.

Now consider the joint ABO in Figure 2-18a. Here we have three

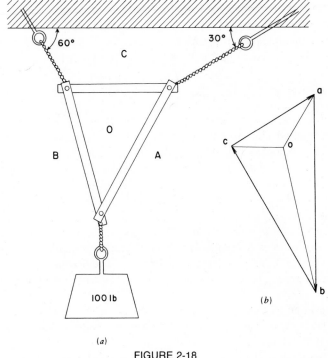

(a)

(b)

FIGURE 2-18

concurrent forces in equilibrium; hence their force polygon must close. Of the three forces only one, *AB*, is known, but we do know the lines of action of *BO* and *OA*. Therefore, in Figure 2-18*b* draw a line through point *b* parallel to *BO* and also a line through point *a* parallel to *OA*. The point *o* must be on both of these lines and therefore is at their point of intersection. Thus the polygon *abo* in Figure 2-18*b* is the force polygon for the three forces *AB*, *BO*, and *OA* in Figure 2-18*a*.

Next consider the joint *BCO*. Again we have three concurrent forces in equilibrium; hence their force polygon must close. Of these three forces, we now know two, *OB* and *BC*. If in Figure 2-18*b* we draw a line connecting points *c* and *o*, *we find that it is parallel to the member CO* and thus checks the accuracy of the forces previously determined. Figure 2-18*b* now contains lines representing the external forces and also the stresses in the members of the triangular frame. We call Figure

2-18*b* a *stress diagram* because it shows the stresses in the members of the frame, *AO*, *BO*, and *CO*. The magnitudes of the stresses in corresponding sides of the frame are established by the lengths of the lines *ao*, *bo*, and *co* in the stress diagram (Figure 2-18*b*).

2-15. Funicular Polygon

It is frequently necessary to determine whether a system of forces is in equilibrium. This may be accomplished by constructing a *funicular polygon*. Suppose that we are given the system of three forces, *AB*, *BC*, and *CA*, shown in Figure 2-19 and are asked to determine whether the system is in equilibrium. The test is to construct the funicular polygon; if both the funicular polygon and the force polygon close, the system is in equilibrium. First, construct the force polygon *abc*, shown in Figure 2-19*b*. The polygon closes. We know that a system of *concurrent* forces is in equilibrium if the force polygon closes, but in this instance let us assume that we do not know whether the lines of action of the forces have a point in common or whether they are nonconcurrent.

In Figure 2-19*b* select *any* point such as point *o* and draw the lines *oa*, *ob*, and *oc*. The point *o* is called the *pole*, and the lines *oa*, *ob* and *oc* are the *rays*. Note that this diagram somewhat resembles Figure 2-18*b* and that the rays *oa*, *ob*, and *oc* represent the stresses in a frame connecting the forces *AB*, *BC*, and *CA*, *if there were such a frame*. Now, let us construct this imaginary frame, the sides of which are sometimes called *strings*. In Figure 2-19*b* the force *ab* is held in equilibrium by the forces *ao* and *bo*. Therefore, select any point on the line of action of force *AB* in Figure 2-19*a* and draw the lines parallel to *oa*

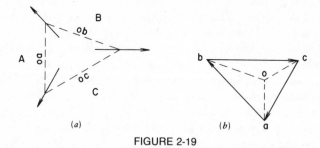

(a)   (b)

FIGURE 2-19

and *ob* in Figure 2-19*b*. These lines represent two sides of the imaginary frame. From the point where *ob* intersects the line of action of the force *BC* draw the line *oc* parallel to the ray *oc*. Note that in Figure 2-19*b* *ob* and *oc* hold the force *bc* in equilibrium and therefore have a point in common. The lines *oa* and *oc* are continued in Figure 2-19*a* until they intersect, and in this example, *they intersect on the line of action of force CA*; we say this imaginary *frame* (*the funicular polygon*) *closes*. The lines in Figure 2-19*a*, *ao*, *ob*, and *oc*, are the sides of an imaginary frame called the *funicular polygon*. This polygon is sometimes called the *string polygon or equilibrium polygon*. Since, in this particular instance, both the force polygon and the funicular polygon close, the three given forces are in equilibrium. In this example note that if the line of action of the force *CA* is extended it will pass through the point of intersection of forces *AB* and *BC*. Actually, the system of three forces is concurrent.

Any system of forces is in equilibrium if both the force polygon and the funicular polygon close.

2-16. Mechanical Couple

The system of three forces shown in Figure 2-20*a* is somewhat similar to the given system shown in Figure 2-19*a*. The respective forces are equal in magnitude, are parallel, and have the same direction, but their lines of action do not bear the same relation to each other; their lines of action do not have a point in common. Is the system of forces shown in Figure 2-20*a* in equilibrium? If it is, both the force polygon and the funicular must close. Let us investigate.

First we draw the force polygon, Figure 2-20*b*, and we find that it

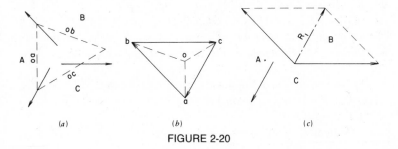

(*a*)   (*b*)   (*c*)

FIGURE 2-20

closes. The next test is the funicular polygon. We select any point, such as *o*, in the force polygon, and draw the rays *oa*, *ob*, and *oc*. Selecting any point on force *AB*, we draw the strings *oa* and *ob* parallel to *oa* and *ob* in the force polygon. At the intersection of string *oa* with force *CA* we draw the string *oc* until it intersects the string *ob*. In this instance we find that the funicular polygon does not close; that is, the strings *oc* and *ob* do not intersect on the line of action of the force *BC*. Consequently, the system is not in equilibrium. The system might be altered to provide equilibrium by moving the force *BC* to a parallel position in which its line of action passes through the intersection of *oc* and *ob*. The system, *thus modified*, would have both the force polygon and the funicular polygon closing. Notice that thus altering the system would result in a system of concurrent forces.

But, if the original three forces shown in Figure 2-20*a* are not in equilibrium, what is their resultant? These three forces are again drawn in Figure 2-20*c*. Let us find the resultant of the two forces *AB* and *BC* using the method explained in Section 2-7. The force polygon so constructed will be similar to Figure 2-20*b* except that *ac*, the resultant, acts upward to the right. (Note that the resultant is equal in magnitude and parallel to *CA*, the third force in the system, but it acts in the opposite direction.)

The resultant of *AB* and *BC* might also have been found by constructing the parallelogram of forces as shown by the dashed lines in Figure 2-20*c*, the resultant being denoted by R_1. Regardless of how it is found, however, the resultant of two *nonparallel* forces passes through the point at which their lines of action intersect.

If, in Figure 2-20*c*, we substitute the force marked R_1 for the two forces *AB* and *BC*, we have remaining the two forces R_1 and *CA*. *These two forces are parallel, equal in magnitude, opposite in direction, and do not have the same lines of action. Such a system constitutes a mechanical couple.* A mechanical couple acting on a body causes motion of rotation rather than motion of translation. A mechanical couple can be held in equilibrium only by the addition of another mechanical couple. It cannot be balanced by a single force.

Thus we have found that a system of forces whose force polygon closes but whose funicular polygon does not close constitutes a mechanical couple.

A mechanical couple is frequently found in daily experiences. A person's two hands operating the steering wheel of an automobile is an example.

2-17. Three Forces in Equilibrium

The lines of action of any system of three nonparallel forces in equilibrium intersect at a common point. For such a system the resultant of any two forces must be the equilibrium of the third force.

Figure 2-19a shows three nonparallel forces in equilibrium. In accordance with Section 2-16, the resultant of *ab* and *bc*, shown in the force polygon, Figure 2-19b, is *ac*, a force whose direction is upward to the right and whose line of action passes through the intersection of *AB* and *BC*. If *ac* is substituted for *ab* and *bc*, we have remaining only *ac* and *ca*, two equal parallel forces opposite in direction. Since by data these forces are in equilibrium, they must have the same line of action. Therefore, *CA* must pass through the point common to *AB* and *BC*.

2-18. Resultant Found by Funicular Polygon

The construction of the funicular polygon affords a convenient method of determining the resultant of any system of forces.

Example.   Let it be required to find the resultant of the four forces, *AB*, *BC*, CD, and *DE* shown in Figure 2-21a. This system may or may not be concurrent.

Solution: The polygon of forces is drawn as shown by the solid lines in Figure 2-21b, and the resultant is *ae*, as explained in Section 2-11. Note that this is the resultant in magnitude and direction only; its line of action is found by the funicular polygon. Point *o*, the pole, is selected, and the rays *oa*, *ob*, *oc*, *od*, and *oe* are drawn. From any point on the line of action of force *BC* we next draw the strings *ob* and *oc* (Figure 2-21a). Care must be exercised in drawing the strings in their proper relation to the various forces. In this particular instance we note that the ray *oc* is common to both *bc* and *cd* in the force polygon, and, therefore, the string *oc* must join the forces *BC* and *CD* in the space diagram (Figure 2-21a). We continue by drawing the strings of the funicular polygon. Now the resultant of the given system

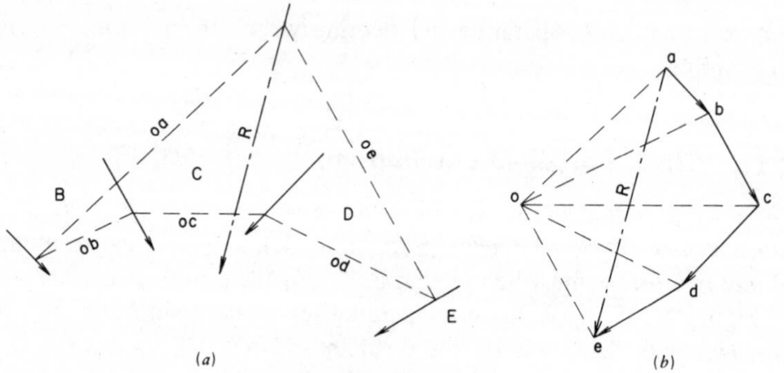

FIGURE 2-21

of forces is found in the force polygon to be *ae*, shown by the dot-and-dash line. This figure also shows that this force *ae* is held in equilibrium by the rays *oa* and *oe*. Therefore, the strings *oa* and *oe* in the funicular polygon are extended until they intersect. This point of intersection is a point on the line of action of the resultant. Therefore we draw the resultant, whose magnitude and direction are found in the force polygon, through the point of intersection of strings *oa* and *oe* and thus completely determine the resultant.

2-19. Resultant of Parallel Forces

The resultant of parallel forces may be found by constructing the funicular polygon.

Example. The five forces shown in Figure 2-22*a* constitute a system of parallel forces. Determine the resultant.
Solution: The force polygon is drawn first, as shown by the solid lines in Figure 2-22*b*. In accordance with the rule given in Section 2-16, the resultant in magnitude and direction is the vertical downward force *af*. Its position with respect to the five forces in the given system is not yet known. A pole is now selected, the rays are drawn, and the funicular polygon is constructed as indicated in Figure 2-22*a*. The resultant *af* in Figure 2-22*b* is held in equilibrium by the rays *oa* and *of*, and, therefore, the intersection of the strings *oa* and *of* in the funicular polygon lies on the line of action of the resultant. The resultant, indicated

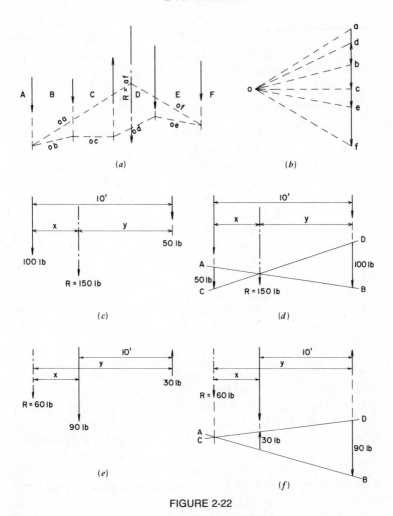

FIGURE 2-22

by the dot-and-dash line, is drawn through this point of intersection and represents the resultant in magnitude, direction, and line of action.

Another graphical method of finding the position of the resultant of parallel forces is the method of *inverse proportion*. Figure 2-22*c* shows two parallel downward forces of 100 and 50 lb. Their resultant is the downward force of 150 lb at x distance from the 100-lb force and y

distance from the 50-lb force. It can be shown by the principle of moments (Section 4-7) that the resultant of the two parallel forces having the same direction divides the distance between them in inverse proportion to the magnitude of the forces; that is, for the forces shown in Figure 2-22c, $100:50::y:x$.

To determine the position of the resultant graphically, assume that the two given forces are 10 ft apart, as shown in Figure 2-22d, and draw any straight line such as AB that intersects their lines of action. On the line of action of the 100-lb force lay off to a suitable scale a downward force of 50 lb from the intersection of AB with the force of 100 lb. Similarly, from the point where AB intersects the 50-lb force lay off a downward force of 100 lb at the same scale *but on the opposite side of the line* AB. From the extremities of the two lengths just laid off draw the line CD. The point of intersection of AB and CD determines the point through which the resultant of the two forces of 100 and 50 lb will pass because, by similar triangles (Figure 2-22d), $100:50::y:x$.

By scaling the lengths of x and y, they are found to be 3.33 and 6.66 ft, respectively. The magnitude of the resultant of two parallel forces having the same direction is equal to their sum, in this instance $50 + 100 = 150$ lb.

Figure 2-22e shows two parallel forces of 90 and 30 lb, 10 ft apart and having opposite directions. Their resultant is a downward force of 60 lb at the position shown. By the principle of moments, $90:30::y:x$. The position of the resultant is found graphically in Figure 2-22f. The line AB is drawn, and it intersects the line of action of the two given forces. We now proceed as before, except that *the forces of 90 and 30 lb are laid off on the same side of the line AB*. This is shown in Figure 2-22f. From the ends of the two forces just laid off we draw the line CD. The point at which AB and CD intersect is a point through which the resultant of the 90- and 30-lb forces will pass. Note that, by similar triangles, $90:30::y:x$. By scaling the distance y, it is found to be 15 ft 0 in. The resultant of two parallel forces having opposite directions is equal to their difference in magnitude and has the direction of the greater force. In this instance the resultant is $90 - 30$, or a downward force of 60 lb.

The preceding discussion explains a graphical method of finding the

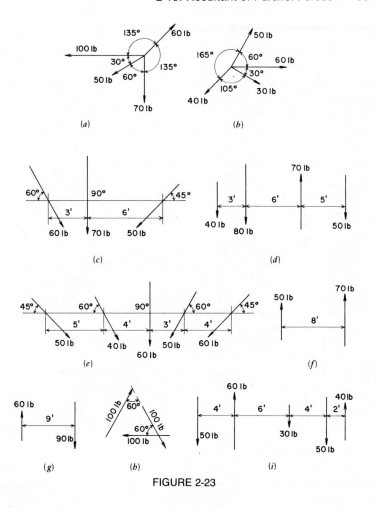

FIGURE 2-23

resultant of two parallel forces. Another method, which is possibly shorter, is to construct a funicular polygon. The shortest and most accurate method, however, is to employ the principle of moments, which is discussed in detail in Chapter 4.

Problems 2-19-A-B*-C-D-E*-F-G-H*-I. Using graphical methods, find the resultants of the systems of forces shown in Figures 2-23a, b, c, d, e, f, g, h, and i.

2-20.   Reactions Found Graphically

The term *reactions* is given to the upward forces or supports that hold a beam in equilibrium. If the loading on a simple beam is symmetrical, the reactions are equal in magnitude. If, however, as frequently happens, the beam is not loaded symmetrically, the magnitudes of the reactions must be computed or found graphically. The usual and most convenient method of determining the reactions is to use the moments principle (Section 4-6), but they may be determined graphically by employing the force and funicular polygons.

Example.   The beam shown in Figure 2-24*a* is 20 ft long and supports three loads of 1000, 800, and 400 lb at the spacings indicated. Determine by graphical methods the magnitudes of the reactions *DE* and *EA*.

Solution: (1) The three loads have been designated *AB, BC,* and *CD.* Since these are vertical downward forces due to gravity, the reactions *DE* and *EA* are vertical upward forces, and the system is composed of five parallel forces in equilibrium. Three of the five forces (the loads) are known completely. The lines of action and directions of the reactions are known but their magnitudes are unknown.

(2) The spacing of the forces is laid off at some convenient scale, in this instance $\frac{1}{8}$ in. = 1 ft 0 in., as shown in the space diagram, Figure 2-24*a*. Then the force polygon is begun by drawing the three known forces, *AB, BC,* and *CD.* This is shown in Figure 2-24*b* the scale of which is 1 in. = 1600 lb. Note that point *e* in the force polygon cannot

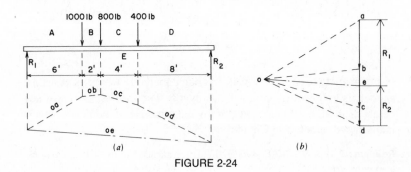

FIGURE 2-24

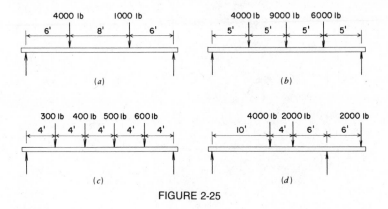

FIGURE 2-25

be located because the magnitudes of *DE* and *EA* are not yet known. A pole is selected, and the rays *oa*, *ob*, *oc*, and *od*, are drawn.

(3) The lines of action of the various forces are continued below the space diagram, and the sides of the funicular polygon, *oa*, *ob*, oc, and *od*, are drawn (Figure 2-24*a*). *Since, by data the five forces are in equilibrium, we know that the funicular polygon must close, and, therefore, we draw the closing string oe.* Now we return to the force polygon, Figure 2-24*b*, and draw the ray *oe* parallel to the string *oe*, thus determining the point *e*. By scaling the lengths of *de* and *ea*, we find them to be 860 and 1340 lb, the magnitudes of the two reactions *DE* and *EA*.

Note that the reactions have been further identified in the force polygon (Figure 2-24*b*) as R_1 and R_2. These convenient designations are widely used to signify the left and right reactions, respectively.

Problems 2-20-A-B-C-D*.   Letter the beams shown in Figures 2-25*a*, *b*, *c*, and *d* in any convenient manner and determine the magnitudes of the reactions by graphical methods.

2-21. Parallel Truss Reactions

For vertical loads placed symmetrically on a roof truss, the reactions at the ends are vertical, and the magnitude of each is equal to half the sum of the loads. Wind loading, however, is not symmetrical, and the

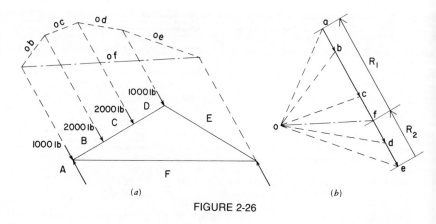

FIGURE 2-26

magnitudes of the reactions due to the wind cannot be found by inspection. The method of determining wind-load reactions presented in this section is based on the assumption that the reactions are parallel to the wind loads.

Figure 2-26*a* shows the upper and lower chords of a truss with the web system omitted to avoid confusion (see Figure 2-28 for terminology). The wind loads are shown acting on the left side of the upper chord, perpendicular to the roof suface, and are designated by the letters *AB*, *BC*, *CD*, and *DE*. The first step is to draw the sides of the force polygon, *ab*, *bc*, *cd*, and *de*, as shown in Figure 2-26*b*. The pole *o* is then selected, and the rays *oa*, *ob*, *oc*, *od*, and *oe* are drawn. Next, the lines of action of the loads and reactions are extended above the truss, and the funicular polygon is constructed (Figure 2-26*a*). Since the six forces are in equilibrium, the funicular polygon must close; therefore, the closing string *of* is drawn. A line from the pole *o* parallel to the closing string determines the point *f* on the force polygon and thus establishes the two reactions *EF* and *FA*. Scaling their lengths shows that they are 1860 and 4140 lb, respectively.

The string *oa* in the funicular polygon requires an explanation. Note that the forces *FA* and *AB* in Figure 2-26*a* have the same line of action and that their resultant is *FB*. If in drawing the funicular polygon only the force *FB* is considered to be acting at the left end of the truss, the string *oa* is unnecessary. Another explanation follows: the string *ob* is extended to the left until it intersects the force *AB*. From this point of intersection *oa* is continued until it intersects the next force *FA*. But *FA* and *AB* have the same line of action, hence the string *oa* has no length. It should be noted, however, that if string *oa* is drawn upward to the right and string *oe* extended upward to the left, their intersection will locate a point on the line of action of the resultant of forces *AB*, *BC*, *CD*, and *DE*.

2-22.   Truss Reactions: Roller at One End

To provide for expansion and contraction that result from temperature changes, a roller is sometimes used at one support of a truss. If such a bearing were frictionless, the reaction at the roller support would always be vertical while the other reaction would be inclined to the right or left depending on the direction of the wind load.

The right end of the truss indicated in Figure 2-27*a* is rollersupported, and the two forces *AB* and *BC* are the resultants of the wind loading and the vertical loading, respectively. To determine the reactions, we first draw the two forces *ab* and *bc* as part of the force polygon (Figure 2-27*b*). The pole *o* is selected, and the rays *oa*, *ob*, and *oc* are drawn. Assuming that the roller at the right end is frictionless, the reaction *CD* (or R_2) is vertical, but the direction of the left reaction is unknown.

The only fact known about the left reaction *DA* is that its line of action must pass through the point of support. Therefore the funicular polygon is started at this point, the string *oa* being continued until it intersects the force *AB*. We continue by drawing the strings *ob* and *oc*,

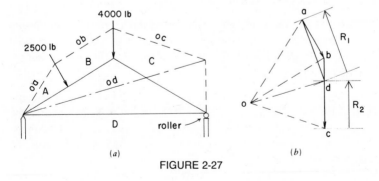

(*a*)   (*b*)

FIGURE 2-27

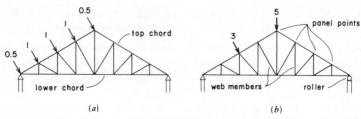

FIGURE 2-28

oc being extended until it meets the line of action of the reaction *CD*, which we know is vertical. From this point of intersection the closing string *od* is drawn to the left support. Now return to the force polygon and draw the ray *od* parallel to the closing string *od*, thus establishing point *d* on the force polygon. The reactions of the truss are *cd* and *da*, their magnitudes and directions being shown on the force polygon.

Problem 2-22-A.   The triangular Howe truss shown in Figure 2-28*a* has a span of 64 ft 0 in., and the upper chord makes an angle of 30° with the horizontal. Assuming that the reactions are parallel to the direction of the wind, draw the force polygon of the external forces (wind loads and reactions) and determine the magnitudes of the reactions.

Problem 2-22-B.   The truss indicated in Figure 2-28*b* is a triangular Howe having a span of 60 ft 0 in., the angle between the upper and lower chords being 30°. The resultant of the wind loads is 3000 lb, and it acts at the mid-length of the left half of the upper chord. The resultant of the vertical loads is 5000 lb at the position shown. Assuming that a roller bearing is used at the right reaction, draw the force polygon of the external forces and scale the magnitudes of the reactions.

2-23.   Truss Reactions: Three Forces in Equilibrium

As stated in Section 2-17, any three nonparallel forces in equilibrium have lines of action that intersect at a common point. This important fact is used frequently in graphical solutions. The following example is an illustration.

Example.   The fan truss shown in Figure 2-29*a* is subjected to the wind loads *AB*, *BC*, *CD*, and *DE*. Roller support is provided at the right end of the truss, while the left support is capable of taking a reaction in any direction. Determine the magnitude and direction of each reaction.

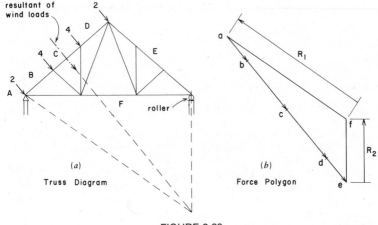

FIGURE 2-29

Solution: (1) We begin by laying off the forces *ab*, *bc*, *cd*, and *de* (sometimes called the *load line*) as part of the force polygon shown in Figure 2-29*b*.

(2) The next force in order is *ef*, a vertical upward force because the roller is located at this reaction. Therefore through point *e* draw a vertical line of indefinite length; point *f* will lie somewhere on this line.

(3) The wind loading is symmetrically distributed on the left side of the truss with the wind resultant acting at mid-length of the slope. Consequently three forces may be considered acting on the truss; the 12,000-lb resultant of the wind loads and the two reactions. *These three forces are in equilibrium and are nonparallel; therefore their lines of action have a point in common.* We now extend the line of action of the wind resultant in the truss diagram until it intersects the vertical line of action of the right reaction. This intersection is the point in common of the three forces, and a line drawn from it to the left support of the truss *determines the line of action of the left reaction.*

(4) With the line of action of the left reaction established, we can draw a line in the force polygon through point *a* parallel to *FA* in the truss diagram. Point *f* lies on this line and also on the vertical line through point *e*. Hence the intersection of these two lines establishes point *f* and consequently determines the magnitudes and direction of the two reactions.

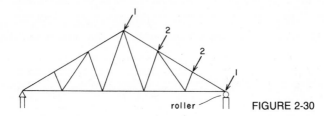

FIGURE 2-30

Problem 2-23-A. The truss indicated in Figure 2-30 is a triangular Belgian truss having a span of 48 ft 0 in. and a height of 16 ft 0 in. The total wind load on the right side of the truss is 6000 lb, distributed at the panel points as shown. A roller is used at the right reaction. Using the method explained in Section 2-23, draw the force polygon of the external forces and determine the magnitudes and directions of the two reactions.

2-24. Stress Diagram for Trusses

Thus far the discussions relating to trusses have been confined to the loads, the reactions, and construction of the force polygon of external forces. The next step is to determine the stresses (internal resisting forces) in the truss members. This may be accomplished by analytic or graphic methods, but only the graphic method of *stress diagram* is considered here.

Figure 2-31*a* represents diagrammatically a triangular fan truss for which the end bearings are simple supports; there is no roller bearing. The total load on the truss is 6000 lb, distributed at the panel points as shown. As the truss is symmetrically loaded, the reactions are vertical and equal in magnitude, each being 3000 lb. Observe that at each panel point there are three or more concurrent forces. At each joint the concurrent forces are in equilibrium, and for this condition we know that their force polygons close. It should be noted that certain forces (the loads and reactions) are known; the remaining forces, the stresses in the truss members, are to be determined. For the unknown forces, note particularly that we do know their lines of action.

Consider first the joint *ABJI* at the left support. The isolated joint diagram in Figure 2-31*b* shows that we have four concurrent forces in equilibrium; the two known forces are *IA* and *AB* (reading clockwise about the joint), and the two unknown forces are *BJ* and *JI*. Begin by drawing, at some convenient scale, the sides of the force polygon *ia* = 3000 lb and *ab* = 500 lb as shown in Figure 2-31*b*. The next force

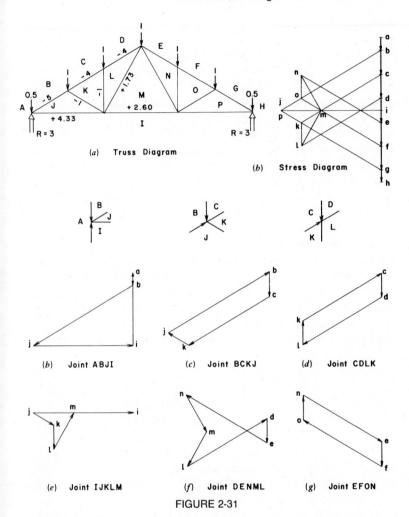

(a)   Truss Diagram

(b)   Stress Diagram

(b)   Joint ABJI

(c)   Joint BCKJ

(d)   Joint CDLK

(e)   Joint IJKLM

(f)   Joint DENML

(g)   Joint EFON

FIGURE 2-31

in order is *BJ*; therefore, through point *b* draw a line parallel to *BJ*. Point *j* will be somewhere on this line. The next force is *JI*; through point *i* draw a line parallel to *JI*. Since point *j* is on this line, as well as on the line through *b* parallel to *BJ*, it must be at their point of intersection. Thus Figure 2-31*b* is the force polygon for the concurrent forces at joint *ABJI*, and we have established the magnitudes of the stresses in the members *BJ* and *JI*, which scale 5000 lb and 4333 lb, respectively.

It is now necessary to determine the *character* of the stress in the members meeting at this joint—that is, whether a member is in compression or tension (Section 2-2 and 2-3). Let us consider member *BJ* first. Because the names of the forces were read in a clockwise direction about the joint, note the sequence of letters: *B* first, then *J*. In the force polygon for the joint, we find that *bj* (read in the same sequence) reads downward to the left. Referring to the isolated joint diagram of Figure 2-31*b* (or to the same joint in the truss diagram of Figure 2-31*a*) and reading *BJ* downward to the left, we read *toward* the center of the joint; hence member *BJ* is in compression. It is important to remember that a member in compression tends to be made shorter and resists this shortening by *pushing against* the joints at its ends. A tension member, on the other hand, tends to become longer and resists the lengthening by *pulling away* from its end joints. In the truss diagram of Figure 2-31*a*, a minus (−) preceding the recorded magnitude denotes compression in the member and a plus (+) denotes tension. This sign convention is widely used but in some books the designations are reversed and in others the symbols (*C*) and (*T*) are employed instead of algebraic signs. It is only necessary to be consistent once a system is adopted.

The other truss member framing into the joint *ABJI* is (again reading clockwise) *JI*. In the force polygon we find that *ji* reads from left to right. Referring to the joint or truss diagram and reading *JI* from left to right, we read *away* from the center of the joint; hence member *JI* is in tension.

The joint *JKLMI* consists of five concurrent forces of which only one, *IJ*, is known. These data are insufficient for constructing the force polygon. *At any joint we can draw the force polygon provided that there are not more than two unknowns.*

Now consider joint *BCKJ*. We have already determined *JB* (read *BJ* in Figure 2-31*b*), and we know that *BC* is 1000 lb. Thus we know two of the forces, and only two forces, *CK* and *KJ*, are unknown. Therefore we know that we can draw a force polygon for the forces at this joint. Begin by drawing *jb* and *bc*, the two known sides of the force polygon, as shown in Figure 2-31*c*. The next force is *CK*; therefore, we draw a line through point *c* parallel to member *CK*. Point *k* will be on this line. The next force is *KJ*; hence through point *j* we draw a line parallel to *KJ*. Point *k* will be on this line, and, consequently, it will be at the point where this line intersects the line through *c* parallel to *CK*. The force polygon now completed is *jbck*, and we have determined the magnitude of two more forces, *CK* and *KJ*. Following the procedure used at joint *ABJI*, the character of the stress in members *CK* and *KJ* is determined, and this information, together with the magnitude of the stresses, is recorded on the truss diagram (Figure 2-31*a*).

In a similar manner the remaining joints are taken in the following sequence: *CDLK*, *IJKLM*, *DENML*, and *EFON*. The reader should sketch isolated joint diagrams relating to Figure 2-31*e*, f, and g. The stresses in the members on the right-hand side of this symmetrically loaded truss are, of course, similar to those for corresponding members on the left.

The determination of stresses in the members of a truss by drawing separate force polygons for each joint is a cumbersome procedure. A more practicable method is to construct a single diagram that combines all the separate force polygons. Such a diagram is called a *stress diagram* because it establishes the stresses (resisting forces) in the members of the truss. It will be observed that each separate force polygon drawn for a joint is also, in effect, a stress diagram. Custom, however, generally reserves the term for the single combined diagram.

To draw a stress diagram we begin by drawing the force polygon of the external forces. For the truss and loads shown in Figure 2-31*a* this diagram consists of the known forces, the loads and reactions. The force polygon is shown at the right in Figure 2-31*h*; it is *ab*, *bc*, *cd*, *de*, *ef*, *fg*, *gh*, *hi*, and *ia*. Now we begin at some point at which there are not more than two unknowns and draw the force polygon in conjunction with the force polygon of external forces previously drawn. First take the joint *ABJI* and construct the polygon *abji* as shown in Figure 2-31*h*. Note that this is exactly the polygon shown in Figure 2-31*b*. The remaining joints are taken as described, and the complete stress diagram for the truss is shown in Figure 2-31*h*.

The magnitudes of the stress (resisting forces) in the various members of the truss are found by scaling the lengths of the lines in the stress diagram. The length of a member is not an indication of the magnitude of the stress in that member. The lengths of the lines *in the stress diagram*, however, do determine the magnitude of the stresses in the truss members.

In the stress diagram method, the character of the stress (compression or tension) in each member is determined by the same procedure as that explained earlier in connection with separate force polygons for each joint. Care must be taken to read the name of the member about the joint in question (called the reference joint) in a clockwise direction, the same direction that was used when drawing the force polygon of external forces in the stress diagram.

The sequence of letters in identifying a member is most important. As an example, consider the vertical truss member that frames into the joint *CDLK* in the truss shown in Figure 2-31*a*. (The forces were read in a *clockwise* manner in drawing the stress diagram for this truss.) This vertical member is read *LK* if the reference joint is *CDLK*, but, if the reference joint is *JKLMI*, the same member is read *KL*. Be sure that you understand this before you continue.

Example.   Determine the character of the 5000-lb stress in the upper chord member adjacent to the left support as shown in Figure 2-31*a*.
Solution: (1) This member is identified as *BJ* (not *JB*) with respect to the reference joint *ABJI*. Note the sequence, the letter *B* coming before *J*.

(2) Turning to the stress diagram (Figure 2-31*h*), we find that this force *bj* (read in the same sequence) reads downward to the left.

(3) Returning to the truss diagram (Figure 2-31*a*), and reading the member *BJ downward to the left* (the direction found in Step 2), we read *toward the reference joint ABJI*; hence the member is in compression.

Let us try again, testing the same member with respect to another reference joint.

(1) Consider the member *JB* with respect to the reference joint *BCKJ*, the first upper-chord panel point on the left side of the truss.

(2) In the stress diagram this force *jb* reads *upward to the right*.

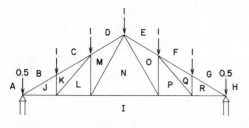

FIGURE 2-32

(3) In the truss diagram we note that, if *JB* has a direction *upward to the right*, we read *toward* the reference joint *BCKJ*; hence the member is in compression. This is, of course, the same character of stress that was found when the other reference joint was used.

Problem 2-24-A*.   The triangular Pratt truss shown in Figure 2-32 has a span of 58 ft 0 in., and the upper chords make angles of 30° with the horizontal. The reactions are vertical. Draw the complete stress diagram for this truss and loading and determine the character and magnitudes of the stresses in the various truss members.

3

Forces Acting on Bodies
IIIIIIIIIIIIIIIIIIIIIIIIIIIIIIIIIIIIIIIIIIIIIIIIII

3-1. Free-Body Diagram

A convenient way to determine the unknown forces acting on a body, or the unknown internal stresses in a structure, is to construct a *free-body diagram*. This may be for a whole structure or a part of a structure. The usual procedure is to imagine one part of the structure to be cut away from the adjoining parts and moved to a free position in space. This isolated object is known as a *free body*. The problems with which we are concerned deal with forces in equilibrium. Hence the forces are represented as vectors (graphically), and the force polygon of the forces acting on the free body must close; this fact enables us to determine the unknown forces.

Consider Figure 3-1*a*, which represents two members framing into a wall, the upper member being horizontal and the angle between them being 30°. A stone block weighing 200 lb is placed over the point at which the two members meet. Figure 3-1*b* is a diagram showing the block as a free body, the three forces being the vertical force of 200 lb (the weight of the block), the unknown force acting along the horizontal member, and the unknown force acting through the inclined member. The system of identifying forces explained in Section 2-14 is used in Figure 3-1*c*; thus the forces acting on the free body are *AB* (the force due to gravity) and the unknowns *BC* and *CA*, although the directions of the arrows on the last two have not yet been determined. It

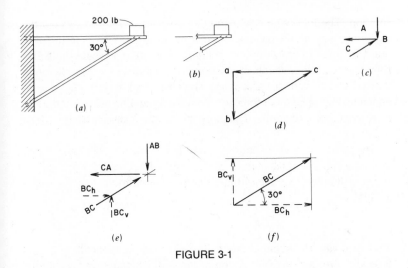

FIGURE 3-1

should be noted that the unknown stresses in the frame members become *external forces* with respect to the free body. Referring to Figure
2-31, it can be observed that the three isolated joint diagrams serve
essentially as free-body diagrams.

To determine the stresses in the frame members, we need only to
construct the force polygon of the concurrent forces. We begin by
drawing the vector (force) *ab*, a downward vertical force of 200 lb, as
shown in Figure 3-1*d*. Any convenient scale of so many pounds to the
inch may be used. The next force in order is *BC*; therefore through
point *b* we draw a line parallel to the force *BC*. Point *c* is at some point
on this line. Next is the force *CA*; hence we draw a line through point
a parallel to the force *CA*. Point *c* is on this line and therefore is at the
point of intersection is at the point of intersection with the line previously drawn through *b* parallel to *BC*. Figure 3-1*d*, then, is the force
polygon for the three forces acting on the block. By scaling the lengths
of the lines in the force polygon, we find that *BC* = 400 lb and *CA* =
346 lb. This is a *graphical solution*, and for such solutions the accuracy
in determining the stresses depends on the accuracy employed in constructing the diagram. Minute accuracy is neither necessary nor desirable, and graphical solutions are generally sufficiently accurate for
practical purposes.

The preceding problem obviously also lends itself to a mathematical solution. The following is an example of such a solution. Consider the free-body diagram of the forces as shown in Figure 3-1e. In this diagram the three forces as shown in Figure 3-1c are represented, but also shown are the components of the force BC in terms of a resolution into vertical and horizontal effects. As shown in Figure 3-1f, these components may replace the force BC completely and be used to represent it in a free-body diagram. The purpose for this is demonstrated in the following work.

The forces in the free-body diagram in this example are constituted as a concentric, coplanar force system. (See Section 2-4.) For such a system, the algebraic conditions for static equilibrium may be stated as follows:

$$\Sigma F_H = 0, \qquad \Sigma F_V = 0$$

That is to say, the summation of the horizontal components of all the forces is zero, and the summation of the vertical components of all the forces is zero. Referring to Figure 3-1e, we now apply these conditions, plus the known geometry of the force vectors (i.e., their directions) to the example problem as follows:

$$\Sigma F_H = 0 = CA + BC_H$$

$$\Sigma F_V = 0 = AB + BC_V$$

To implement these algebraically, we must assume a sign convention (+ and −) for the force vectors, as follows:

$$+ = \uparrow, \quad \text{and} \quad - = \downarrow$$

$$+ = \rightarrow, \quad \text{and} \quad - = \leftarrow$$

We then proceed by first using the equilibrium equation that contains only one unknown.

$$\Sigma F_V = 0 = -200 + BC_V, \qquad BC_V = +200 \quad \text{or} \quad 200\uparrow$$

Then, from the geometry of BC;

$$BC_V = BC \, (\sin 30°) = BC \, (0.50)$$

Thus

$$BC_V = 200 = BC \, (0.50), \qquad BC = 400 \text{ lb compression}$$

Note that the sense of BC, being in compression, is evident from the sign of BC_V and inspection of the free-body diagram (Figure 3-1e).

We now proceed to use the other equilibrium equation to find the unknown force CA.

$$\Sigma F_H = 0 = CA + BC_H$$

$$= CA + (BC \times \cos 30°)$$

$$= CA = 0.866\,BC$$

$$= CA + 346$$

Thus

$$CA = -346 \quad \text{or} \quad 346 \leftarrow$$

Again, we note that this indicates tension in the frame member from observation of the free-body diagram.

Additional algebraic solutions are illustrated in subsequent chapters; however, in this chapter the remaining examples utilize only graphical solutions.

3-2. Two-Force Members

When a member in equilibrium is acted on by forces at only *two points*, it is known as a *two-force member*. The resultant of all the forces at one point must be equal, opposite in direction, and have the same line of action as the resultant of the forces acting at the other point. The stress in a two-force member is either tension or compression. A vertical post with a load on its upper end is an illustration. Neglecting the weight of the post, the load on the top (due to gravity) is equal in magnitude to the upward reaction at the base of the post, opposite in direction, and has the same line of action. The stress within the post is *axial*.

In Figure 3-1a each of the two members of the frame is a two-force member. Consider, for example, the horizontal member. The forces acting on the right-hand end are the vertical load of 200 lb and the compressive force from the lower frame member. From Figure 3-1d it is seen that the *resultant* of these two forces is *ac*. The force *ac* is horizontal, and the reaction at the wall (the other force acting on this member) is also horizontal, equal to *ac* in magnitude, and with the same line of action.

A roof truss is a framed structure in which the members are framed together to form triangles. In determining the stresses in the truss members, it is customary to ignore the weights of the members since they are small in comparison to the loads. Likewise, we assume that the members are not restrained at their ends; theoretically, they are joined together by pins. With these assumptions, the truss members are two-force members and resist either compressive or tensile forces.

3-3. Three-Force Members

When forces act at three points on a member, it is called a *three-force member*, and the stresses in the member are not axial.

Consider Figure 3-2*a* in which a horizontal bar 6 ft 0 in. in length is pinned to a wall and its left-hand end is supported by a rod attached to the wall 6 ft 0 in. above the bar. At 4 ft 0 in. from the right end of the bar a load of 1000 lb is suspended.

The rod, which is subjected to tensile stresses, is a two-force member; therefore its stress is axial, and the wall must offer a resisting tensile force having the same line of action as the force resisted by the rod. The horizontal bar is a three-forced member being subjected to a force at each end and to a vertical load of 1000 lb at 2 ft 0 in. from

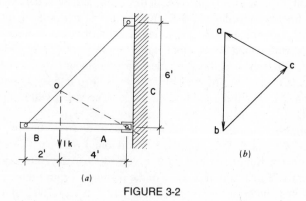

FIGURE 3-2

the left end. The stresses in the bar are not axial; this member is subjected to forces that produce bending stresses, discussed in Section 11-1.

The frame is subjected to three external forces which are identified as *AB*, the vertical force of 1000 lb, *BC*, the force resisting the pull from the rod at the wall, and *CA*, the resisting thrust from the bar at the pinned end. Thus far the direction of the last resisting force is unknown. However, we know that the three external forces are in equilibrium, and, since they are not parallel, they must be concurrent. To find the point at which the three forces meet, extend upward the line of action of the vertical force of 1000 lb and note where it intersects the tie rod, point *o*. From this point draw a line to the pinned end of the bar and thus establish the direction of the thrust at the wall. The directions of all three external forces are now known.

To find the magnitudes of the two resisting forces, construct the force polygon. At a convenient scale draw *ab*, the vertical force of 1000 lb (Figure 3-2*b*). The next force is *BC*; therefore, through point *b* draw a line parallel to the rod *BC*. Next draw a line through point *a* parallel to the direction of the thrust at the pinned end of the bar. The intersection of this line through *b* parallel to *BC* establishes point *c*, thus completing the force polygon. By scaling the length of the lines *bc* and *ca*, we find them to be 943 and 745 lb, respectively.

3-4. The Inclined Plane

When a body acts on another body having a perfectly smooth surface, the body having the smooth surface can exert only a perpendicular force to the contacting surfaces. If the surface on which the body acts is not a theoretically smooth surface, the surface may exert forces inclined to the contacting surfaces. *Friction* is the resistance to sliding between contacting surfaces when the resistance is due to the nature of the surfaces and not to their shape or form.

Figure 3-3*a* represents an inclined plane making an angle of 25° with the horizontal. On the inclined plane is a block of stone weighing 500 lb. Let us assume that the block does not slide, that the friction between the block and the surface of the plane is sufficient to prevent motion. The two forces that hold the vertical force of 500 lb in equilibrium are the resisting pressure from the plane, whose direction is

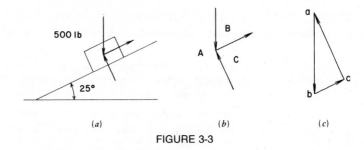

(a)   (b)   (c)

FIGURE 3-3

perpendicular to the plane, and the force parallel to the plane that is the result of friction.

A particle of the block is taken as a free body, and the three forces producing equilibrium are drawn as shown in Figure 3-3*b*. These forces are identified as *AB*, the vertical force of 500 lb, *BC*, the frictional component, and *CA*, the normal component. The forces are concurrent; *AB* is known, but of *BC* and *CA* only the directions are known. As the forces are in equilibrium, their force polygon will close. To construct this polygon, draw, at any convenient scale, a vertical line *ab* having a length equivalent to 500 lb, as shown in Figure 3-3*c*. The next force is *BC*; therefore, through point *b* draw a line parallel to force *BC*. Next draw a line through point *a* parallel to force *CA*. Point *c* is at the intersection of these two lines, and the force polygon representing equilibrium is completed. By scaling the lengths of lines *bc* and *ca* in Figure 3-3*c*, we find *BC*, the force due to friction, to be 211 lb, and *CA*, the normal pressure on the inclined plane, to be 453 lb.

Forces on objects on inclined planes are discussed further in Chapter 19 together with a more complete discussion of the subject of friction.

3-5.   Forces Exerted by Spheres

Figure 3-4*a* indicates a sphere weighing 400 lb which rests in a trough composed of two flat *smooth* surfaces. The surfaces of the trough make angles of 45° and 30° with the horizontal as shown. What forces are exerted on the sides of the trough by the sphere?

Since by data the surfaces of the trough are smooth, the sphere can exert only perpendicular forces to the surfaces of the trough. These

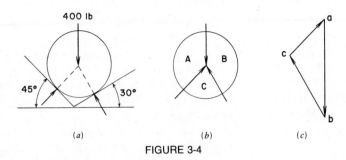

FIGURE 3-4

resisting forces are shown in Figure 3-4a; their lines of action must pass through the center of the sphere. Now that their *directions* are known, their magnitudes are readily found. Figure 3-4b is a free-body diagram of the forces acting on the sphere. AB is 400 lb, the weight of the sphere, and BC and CA are the resisting forces that hold it in equilibrium. The force polygon abc, Figure 3-4c, is constructed as previously described, and, by scaling the lengths of the lines BC and CA, the resisting forces exerted by the surfaces of the trough are found to be 295 and 205 lb, respectively.

Example. Two spheres of unequal size, whose radii are 9 and 6 in., rest in a trough having smooth sides as shown in Figure 3-5a. If the larger sphere weighs 400 lb and the smaller sphere weighs 250 lb, what pressures are exerted at points P, Q, R, and S?

Solution: (1) A free-body diagram of the smaller sphere is drawn as indicated in Figure 3-5b, the pressure at point P having its line of

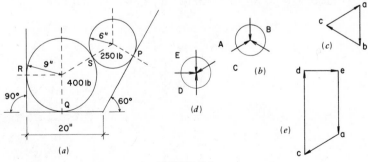

FIGURE 3-5

action perpendicular to the side of the trough and passing through the center of the sphere; the line of action of the force exerted on the larger sphere coincides with the line connecting the centers of the two spheres. The three forces are identified as *AB*, *BC*, and *CA*, *AB* being a vertical force of 250 lb; *BC* and *CA* are of unknown magnitudes. These three concurrent forces are in equilibrium; hence their force polygon will close. We begin by drawing force *ab* at a convenient scale as shown in Figure 3-5*c*. The lines *bc* and *ca* are now constructed as previously explained, and, by scaling, their lengths are found to be 273 and 260 lb, respectively.

(2) Figure 3-5*d* is the free-body diagram of the larger sphere. The force *EA* is 400 lb, the weight of the sphere; *AC* is 260 lb, the thrust from the smaller sphere; *CD* and *DE* are the resisting forces at points *Q* and *R*. Of these four forces, *EA* and *AC* are known; of *CD* and *DE*, only their directions are established. To draw the force polygon, we draw *ea* and *ac* and finally *cd* and *de* (Figure 3-5*e*). By scaling the lengths of the sides of the polygon, *cd* is found to be 515 lb and *de*, 235 lb. Thus the pressures exerted at points *P*, *Q*, *R*, and *S* are 273, 515, 235, and 260 lb, respectively.

3-6.   Force Required to Produce Motion

A wheel 12 in. in diameter weighs 400 lb. Determine the magnitude of the horizontal force, shown in Figure 3-6*a*, required to start the wheel over the 3-in.-high block.

The force necessary to produce motion must be a force slightly greater in magnitude than the force required for equilibrium. The free-body diagram is drawn as shown in Figure 3-6*b*, the three forces being

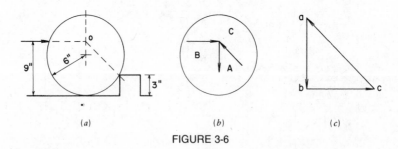

(a)   *(b)*   *(c)*

FIGURE 3-6

the vertical force of 400 lb acting through the center of the wheel, the horizontal force whose line of action intersects the vertical force at point o, and the force whose line of action passes through point o and the point of contact of the wheel and the block. These forces are identified as AB, BC, and CA. The line of action of the last force is established because three nonparallel forces in equilibrium must be concurrent; that is, they must have a point in common. To draw the force polygon, begin by laying off at a suitable scale the vertical force ab, 400 lb, as shown in Figure 3-6c. Then draw a line through b parallel to BC and a line through a parallel to CA. Their intersection determines the point c, and by scaling, we find the horizontal force BC to be 330 lb. A force in excess of 330 lb is required to start the wheel over the block.

Example. Figure 3-7a represents a masonry pier that weighs 10,000 lb. Determine the magnitude of the horizontal force applied at the upper left edge that will be required to overturn the pier.

Solution: The free-body diagram (Figure 3-7b) shows the three forces acting on the pier—AB, the vertical force of 10,000 lb; BC, the horizontal force of unknown magnitude whose line of action intersects the force AB at point o; and CA, the equilibrant of AB and BC, whose line of action passes through point o and the lower right-hand corner of the pier about which the pier will overturn. The line of action of CA is determined by the knowledge that three forces that are in equilibrium and not parallel must have a point in common. Since the lines of action of the three forces are known, and also the magnitude of one of the forces, the force polygon is constructed (Figure 3-7c). By scaling, the

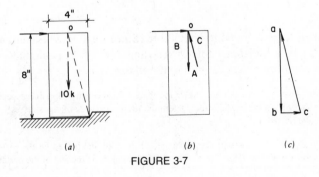

(*a*)   (*b*)   (*c*)

FIGURE 3-7

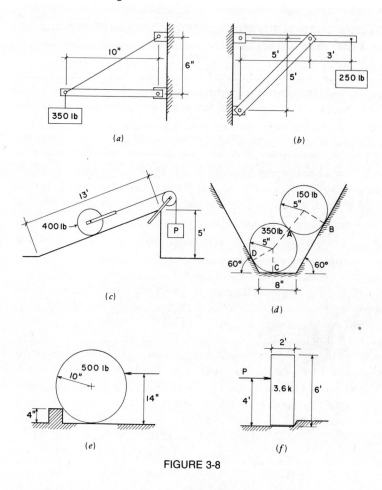

FIGURE 3-8

horizontal force is found to be 2500 lb. Since these three forces are in equilibrium, a horizontal force exceeding 2500 lb in magnitude will disturb equilibrium and cause the pier to turn over.

Problem 3-6-A.   In the structure shown in Figure 3-8*a* a weight of 350 lb is suspended from the end of the horizontal strut. Determine the magnitudes of the stresses in the strut and the tie rod.

Problem 3-6-B.   A horizontal bar is hinged at the wall and braced by the diagonal strut as shown in Figure 3-8*b*. For a load of 250 lb hung from the end of the bar determine the magnitudes of the stress in the strut and the reaction at the hinged joint of the bar.

Problem 3-6-C.   A roller weighing 400 lb rests on a smooth inclined plane having the dimensions shown in Figure 3-8c. Determine the magnitude of the load P required to prevent the roller from moving.

Problem 3-6-D*.   Two cylinders of equal size, weighing 350 and 150 lb, respectively, rest in the trough shown in Figure 3-8d. Determine the pressures at points A, B, C, and D.

Problem 3-6-E.   A cylinder with a diameter of 20 in. weighs 500 lb. What horizontal force P will be required to start the cylinder over the block shown in Figure 3-8e?

Problem 3-6-F.   The pier of masonry indicated in Figure 3-8f weighs 3600 lb. If the horizontal force P is 800 lb, will the pier turn over? What magnitude of force would be required to cause overturning?

4

Moments of Forces

IIIIIIIIIIIIIIIIIIIIIIIIIIIIIIIIIIIIIIIIIIIIIIIII

4-1. Moment of a Force

The term *moment of a force* is commonly used in engineering problems. *A moment is the tendency of a force to cause rotation about a given point or axis.* The magnitude of the moment of a force about a given point or axis is the magnitude of the force (pounds, kips, tons, etc.) multiplied by the perpendicular distance (inches, feet, etc.) between the line of action of the force and the given point or axis. The product of multiplying pounds by feet is neither pounds nor feet, it is their combination, foot-pounds. The point or axis about which the force tends to cause turning is called the *center of moments*. The perpendicular distance between the line of action of the force and the center of moments is called the *lever arm* or *moment arm*. Thus

moment of force = magnitude of force × moment arm

Consider the horizontal force of 100 lb shown in Figure 4-1. If point *A* is the center of moments, the lever arm of the force is 5 ft 0 in. Then the moment of the 100-lb force with respect to point *A* is 100 × 5, or 500 ft-lb. In this illustration the force tends to cause a *clockwise* rotation (shown by the dotted arrow) about point *A* and is called a positive moment. Since 1 ft 0 in. = 12 in., we may multiply 500 ft-lb by 12; the product is 6000 in-lb, a moment of the same magnitude.

In the same figure the 100-lb force has a lever arm of 3 ft 0 in. with respect to point *B*. Therefore the moment of the 100-lb force about

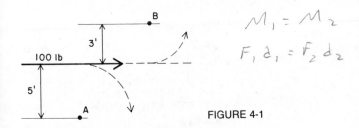

$$M_1 = M_2$$
$$F_1 d_1 = F_2 d_2$$

FIGURE 4-1

point *B* is 100 × 3, or 300 ft-lb, which equals 3600 in-lb. With respect to point *B*, the force tends to cause *counterclockwise* rotation; we call it a negative moment.

It is important to remember that in considering the moment of a force we must have definitely in mind the specific point or axis about which the force tends to cause rotation. Later in this book we write *equations of moments* in which more than one moment is considered. In such an equation we must be certain that *the same center of moments* is taken for all the various moments.

4-2. Increasing Moments

A moment may be increased by increasing the magnitude of the force or by increasing the length of the lever arm.

Figure 4-2 represents a wrench used to turn nuts on bolts. Assume that a nut is screwed up tight. A vertical force of 50 lb is exerted 10 in. from the center of the bolt, thus producing a moment of 50 × 10, or 500 in-lb. It is found that this moment is insufficient to produce motion. A length of pipe is fitted over the handle of the wrench so that the lever arm is increased to 25 in. With the same force of 50 lb, the moment is now 50 × 25, or 1250 in-lb, and motion results. Thus the moment has been increased by increasing the length of the lever arm.

Suppose that we are asked to determine the magnitude of the force necessary to produce motion if the lever arm remains 10 in. Let us call the unknown force *P* pounds. It was found that the moment required to produce motion was 1250 in-lb. Then

$$P \times 10 = 1250 \quad \text{and} \quad P = 125 \text{ lb}$$

which is the force required to produce motion if the lever arm is 10 in. long.

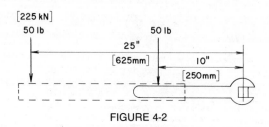

FIGURE 4-2

4-3.   Moment of a Mechanical Couple

In Section 2-18 we learned that a *mechanical couple* consists of two parallel forces equal in magnitude, opposite in direction, and not having the same line of action. A mechanical couple is shown in Figure 4-3; the two equal forces are P_1 and P_2, and the perpendicular distance between their lines of action is x feet.

If we take point a, any point on the line of action of P_2, as the center of moments, the sum of the moments of the two forces about this point is $(P_1 \times x) + (P_2 \times 0)$, which, of course, equals $P_1 \times x$. The moment of P_2 about point a is $(P_2 \times 0)$ because the lever arm is zero. Therefore, the *moment of a couple* is the product of one of the forces and the perpendicular distance between their lines of action.

A mechanical couple produces motion of rotation; it can be held in equilibrium only by the addition of another mechanical couple. The direction of rotation for the couple shown in Figure 4-3 is counterclockwise.

4-4.   Moments of Forces on a Beam

Figure 4-4a shows a cantilever beam with a concentrated load of 100 lb [445 N] placed 4 ft [1.22 m] from the face of the supporting wall. In this position the moment of the force about point A is $4 \times 100 =$

FIGURE 4-3

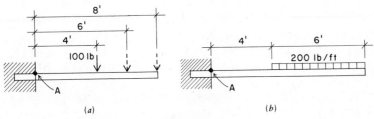

FIGURE 4-4

400 ft-lb [445 × 1.22 = 542.9 N-m]. If the load is moved 2 ft [0.61 m] farther to the right, the moment of the force about point *A* is 600 ft-lb [814.4 N-m]. When the load is moved to the end of the beam the moment about the same point *A* is 800 ft-lb [1085.8 N-m].

Figure 4-4*b* shows a cantilever beam 10 ft [3.048 m] in length with a uniformly distributed load of 200 lb per ft [2.92 kN/m] extending over a length of 6 ft [1.83 m] at the position indicated. Because w = 200 lb per ft, W (the total distributed load) = 200 × 6 = 1200 lb [2.92 × 1.83 = 5.34 kN]. In computing the moment of a distributed load about some specific point, we consider that the load is acting at its center of gravity. For the distributed load in Figure 4-4*b* the center of gravity is 3 ft [0.914 m] from the right end of the beam. Suppose we are asked to find the moment of this load about point *A* at the face of the wall. With the total load acting at its center of gravity the moment is 1200 × 7 = 8400 ft-lb [5.34 × 2.134 = 11.40 kN-m]. The distance 7 ft [2.134 m] is called the *lever arm* or *moment arm*. Point *A* is sometimes referred to as the *center of moments*. The important thing to remember is that in computing a moment we must have in mind the particular point about which the moment is taken.

In Section 4-5 we write an *equation of moments* in which moments of several forces occur. In writing such an equation, the moment of each force must be taken *about the same point*. Careful attention to identifying the center of moments and the lengths of the moment arms of the forces involved will lead to a ready understanding of the concept of moments, which is fundamental to much of structural analysis and design. It is also essential to a precise understanding of moments that their values be expressed in the correct units; if one does not know whether a moment magnitude is foot-pounds or inch-pounds, one does not know the value of the moment.

4-5.   Laws of Equilibrium

When a body is acted on by a number of forces each force tends to move the body. If the forces are of such magnitude and position that their combined effect produces no motion of the body, the forces are said to be in *equilibrium*.

For coplanar forces the three fundamental laws of equilibrium follow:

1.   The algebraic sum of the horizontal forces equals zero.
2.   The algebraic sum of all the vertical forces equals zero.
3.   The algebraic sum of the moments of all the forces about any point equals zero.

Figure 4-5*a* represents a body subjected to two equal horizontal forces with the same line of action. Obviously the forces are in equilibrium, one balancing the other. Call the force acting toward the right positive and the force acting toward the left negative. Then, in accordance with the first law of equilibrium,

$$500 - 500 = 0 \quad \text{or} \quad 500 \text{ lb} = 500 \text{ lb}$$

Figure 4-5*b* is a simple beam. Four vertical forces act on this beam and are in equilibrium. The two downward forces, or loads, are 4000 and 8000 lb [17.8 and 35.6 kN]. The two upward forces, or reactions, are 4400 and 7600 lb [19.57 and 33.80 kN]. If we call the upward

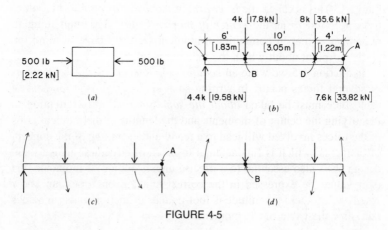

FIGURE 4-5

forces positive and the downward forces negative, by the second law of equilibrium

$$4400 + 7600 - 4000 - 8000 = 0$$

$$[17.8 + 35.6 - 19.58 - 33.82 = 0]$$

Accordingly, if the forces are in equilibrium, we may say that the sum of the downward forces equals the sum of the upward forces or that the sum of the loads equals the sum of the reactions. If, for instance, we had known the magnitudes of the loads and also the magnitude of the left reaction, we could have written

$$4000 + 8000 = 4400 + R_2$$

$$[17.8 + 35.6 = 19.58 + R_2]$$

which yields $R_2 = 7600$ lb [33.82 kN].

Now consider the third law of equilibrium. Take the point marked A on the line of action of the right support and write an equation of moments. (See Figure 4-5c.) Consider moments that tend to cause clockwise rotation as positive and those that tend to cause counterclockwise rotation as negative. The directions of the rotations are indicated by the curved arrow lines. then

$$(4400 \times 20) - (4000 \times 14) - (8000 \times 4) = 0$$

or

$$(4400 \times 20) = (4000 \times 14) + (8000 \times 4)$$

or

$$88,000 \text{ ft-lb} = 88,000 \text{ ft-lb}$$

$$[(19.58 \times 6.10) = (17.8 \times 4.27) + (35.6 \times 1.22)]$$

$$[119.4 \text{ kN-m} = 119.4 \text{ kN-m}]$$

It is now apparent that in accordance with the third law of equilibrium the sum of the moments that tend to cause clockwise rotation equals the sum of the moments that tend to cause counterclockwise rotation.

Note carefully that in writing the equation of moments the moment of each force is taken with respect to the same point, point A in this case. Attention is also called to the moment of the 7600 lb [33.82 kN] force about point A. Because the line of action of this force passes through point A, it can cause no rotation about the point. The lever arm of the moment is zero, or $7600 \times 0 = 0$. We may say that the moment of a force about a point in its line of action is zero. When writing an equation of moments, then, it will be unnecessary to consider the moment of a force if the center of moments is on the line of action of the force.

According to the third law of equilibrium, we may consider moments about *any* point. Let us take point B in Figure 4-5d and see if the law holds. Then

$$(4400 \times 6) + (8000 \times 10) = (7600 \times 14)$$

$$26{,}400 + 80{,}000 = 106{,}400$$

$$106{,}400 \text{ ft-lb} = 106{,}400 \text{ ft-lb}$$

$$[(19.58 \times 1.83) + (35.6 \times 3.05) = (33.82 \times 4.27)]$$

$$[35.83 + 108.58 = 144.41]$$

$$[144.41 \text{ kN-m} = 144.41 \text{ kN-m}]$$

Here again we have omitted writing the moment of the 4000-lb [17.8 kN] force about point B, for we know that its lever arm is zero.

Another example that will serve to fix in mind the principle of moments is given in Figure 4-6. The beam is 9 ft [2.74 m] long with a single pedestal-type support located 6 ft [1.83 m] from the left end. What should the magnitude of the load be at the right end of the beam to produce equilibrium? Call this unknown force x and consider the point of support as the center of moments. Then, because the moment of the force that tends to cause clockwise rotation must equal the mo-

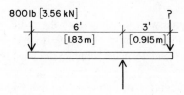

FIGURE 4-6

ment of the force that tends to cause counterclockwise rotation, we may write

$$3 \times x = 800 \times 6 \quad \text{or} \quad 3x = 4800 \quad \text{and} \quad x = 1600 \text{ lb}$$

$$[0.915 \times x = 3.56 \times 1.83 \quad \text{or} \quad 0.915x$$

$$= 6.515 \quad \text{and} \quad x = 7.12 \text{ kN}]$$

The reaction supplied by the single support is $800 + 1600$ or 2400 lb [$3.56 + 7.12$ or 10.68 kN].

Problem 4.5.A. Write the two equations of moments for the four forces shown in Figure 4-5b, taking points C and D as the centers of moments.

4-6. Determination of Reactions

As noted earlier, *reactions* are the forces at the supports of beams that hold the loads in equilibrium. We have seen that the sum of the loads is equal to the sum of the reactions. It is the third law of equilibrium that enables us to compute the magnitudes of the reactions. Throughout this book the left reaction is denoted R_1 and the right reaction is denoted R_2. Another designation frequently used is R_L and R_R.

A simple beam 12 ft [3.655 m] in length has a concentrated load of 1800 lb [8.00 kN] at a point 9 ft [2.74 m] from the left support, as shown in Figure 4-7. Let us compute the reactions. To do this we simply write the sum of moments that tend to cause clockwise rotation and equate it to the sum of the moments that tend to cause counterclockwise rotation. Then, taking R_2 (the right reaction) as the center of moments,

$$12 \times R_1 = 1800 \times 3 \quad \text{or} \quad 12R_1 = 5400 \quad \text{and} \quad R_1 = 450 \text{ lb}$$

$$[3.66 \times R_1 = 8.00 \times 0.915 \quad \text{or}$$

$$3.655R_1 = 7.32 \quad \text{and} \quad R_1 = 2.0 \text{ kN}]$$

To find R_2 we know that $R_1 + R_2$ equals the load on the beam. Then

$$R_1 + R_2 = 1800 \quad \text{or} \quad 450 + R_2 = 1800$$

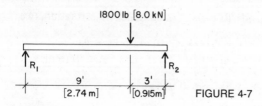

FIGURE 4-7

and

$$R_2 = 1800 - 450 = 1350 \text{ lb}$$

$$[R_2 = 8.0 - 2.0 = 6.0 \text{ kN}]$$

To see whether we have made an error we can check the magnitude of R_2 by writing an equation of moments about R_1 as the center of moments. Then

$$12 \times R_2 = 1800 \times 9 \quad \text{or} \quad 12R_2 = 16{,}200 \quad \text{and} \quad R_2 = 1350 \text{ lb}$$

Note: For the SI computation we use distance values with more digits of accuracy to verify the reactions.

$$[3.6576 \times R_2 = 8.0 \times 2.7432 \quad \text{or} \quad 3.6576R_2 = 21.9456]$$

$$\left[R_2 = \frac{21.9456}{3.6576} = 6.0 \text{ kN} \right]$$

Example 1. The simple beam in Figure 4-8 is 15 ft [4.5 m] long and has three concentrated loads, as indicated. Compute the reactions.

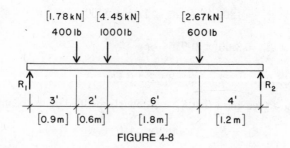

FIGURE 4-8

Solution: Taking R_2 as the center of moments,

$$15R_1 = (400 \times 12) + (1000 \times 10) + (600 \times 4)$$

$$15R_1 = 17{,}200 \quad \text{and} \quad R_1 = 1146.7 \text{ lb}$$

$$[4.5R_1 = (1.78 \times 3.6) + (4.45 \times 3.0) + (2.67 \times 1.2)]$$

$$[4.5R_1 = 22.962 \quad \text{and} \quad R_1 = 5.10 \text{ kN}]$$

To compute R_2 take R_1 as the center of moments. Then

$$15R_2 = (400 \times 3) + (1000 \times 5) + (600 \times 11)$$

$$15R_2 = 12{,}800 \quad \text{and} \quad R_2 = 853.3 \text{ lb}$$

$$[4.5R_2 = (1.78 \times 0.9) + (4.45 \times 1.5) + (2.67 \times 3.3)]$$

$$[4.5R_2 = 17.088 \quad \text{and} \quad R_2 = 3.80 \text{ kN}]$$

To check the results

$$400 + 1000 + 600 = 1146.7 + 853.3$$

$$2000 \text{ lb} = 2000 \text{ lb}$$

$$[1.78 + 4.45 + 2.67 = 5.10 + 3.80]$$

$$[8.90 \text{ kN} = 8.90 \text{ kN}]$$

Example 2.   Compute the reactions for the overhanging beam shown in Figure 4-9.

Solution: Taking R_2 as the center of moments, we equate the clockwise moments to the counterclockwise moments:

$$18R_1 + (600 \times 2) = (200 \times 22) + (1000 \times 10) + (800 \times 4)$$

$$18R_1 = 16{,}400 \quad \text{and} \quad R_1 = 911.1 \text{ lb}$$

$$[5.4R_1 + (2.67 \times 0.6) = (0.89 \times 6.6) + (4.45 \times 3.0)$$

$$+ (3.56 \times 1.2)]$$

$$[5.4R_1 = 21.894 \quad \text{and} \quad R_1 = 4.05 \text{ kN}]$$

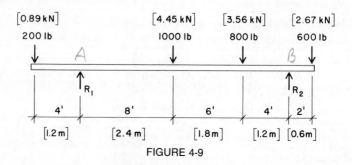

FIGURE 4-9

Taking R_1 as the center of moments,

$$18R_2 + (200 \times 4) = (1000 \times 8) + (800 \times 14) + (600 \times 20)$$

$$18R_2 = 30,400 \quad \text{and} \quad R_2 = 1688.9 \text{ lb}$$

$$[5.4R_2 + (0.89 \times 1.2) = (4.45 \times 2.4) + (3.56 \times 4.2)$$

$$+ (2.67 \times 6.0)]$$

$$[5.4R_2 = 40.584 \quad \text{and} \quad R_2 = 7.52 \text{ kN}]$$

Check

$$200 + 1000 + 800 + 600 = 911.1 + 1688.9$$

$$2600 \text{ lb} = 2600 \text{ lb}$$

$$[0.89 + 4.45 + 3.56 + 2.67 = 4.05 + 7.52]$$

$$[11.57 \text{ kN} = 11.57 \text{ kN}]$$

Example 3. The simple beam shown in Figure 4-10 has one concentrated load and a uniformly distributed load over part of its length. Compute the reactions.

Solution: The total uniformly distributed load is $200 \times 8 = 1600$ lb [$2.92 \times 2.4 = 7.0$ kN]. The center of gravity of this load is at the center of the 8 ft [2.4 m] distance; therefore, insofar as the determination of the reactions is concerned, this loading is the same as that shown in Figure 4-10b. The equation of moments with R_2 as the center

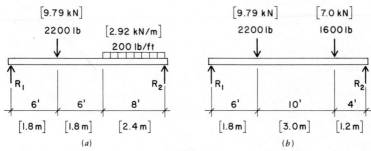

FIGURE 4-10

of moments is

$$20R_1 = (2200 \times 14) + (200 \times 8 \times 4)$$
$$20R_1 = 37{,}200 \quad \text{and} \quad R_1 = 1860 \text{ lb}$$

$$6.0R_1 = (9.79 \times 4.2) + (2.92 \times 2.4 \times 1.2)$$
$$6.0R_1 = 49.5276 \quad \text{and} \quad R_1 = 8.25 \text{ kN}$$

Taking R_1 as the center of moments,

$$20R_2 = (2200 \times 6) + (200 \times 8 \times 16)$$
$$20R_2 = 38{,}800 \quad \text{and} \quad R_2 = 1940 \text{ lb}$$

$$\left[\begin{array}{c} 6.0R_2 = (9.79 \times 1.8) + (2.92 \times 2.4 \times 4.8) \\ 6.0R_2 = 51.2604 \quad \text{and} \quad R_2 = 8.54 \text{ kN} \end{array} \right]$$

Check

$$2200 + 1600 = 1860 + 1940$$
$$3800 \text{ lb} = 3800 \text{ lb}$$

$$\left[\begin{array}{c} 9.79 + 7.0 = 8.25 + 8.54 \\ 16.79 \text{ kN} = 16.79 \text{ kN} \end{array} \right]$$

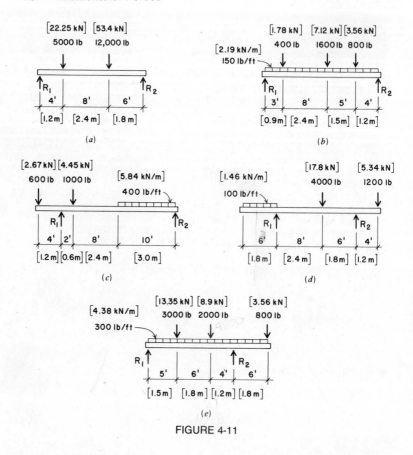

FIGURE 4-11

Problems 4-6-A-B-C*-D-E.* Compute the reactions for the beams shown in Figure 4-11a, b, c, d, and e.

4-7. Forces Found by Moments

The principle of moments is employed constantly in everyday problems, and its use enables us to determine the magnitude of unknown forces. For many of the illustrative beam problems previously given the weight of the beam has been ignored; this was done purposely to simplify the explanations. Actually the weight of the beam constitutes a uniformly distributed load and, with respect to the reactions, is con-

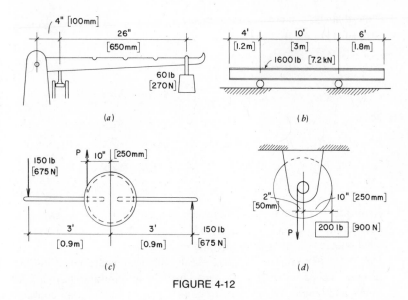

FIGURE 4-12

sidered to act at its center of gravity. In the following examples the weights of the members as well as the resistance due to friction is purposely omitted.

Example 1. A safety valve and lever are indicated in Figure 4-12a. What force due to pressure would be required to open the valve if the suspended weight were 60 lb [270 N]?

Solution: Taking the hinge as the center of moments, we may write

$$60 \times 30 = \text{pressure} \times 4$$

$$\text{pressure} = 450 \text{ lb } [2025 \text{ N}]$$

If the weight of the lever arm and friction were ignored, a valve pressure of 450 lb would balance the 60 lb weight. Theoretically, any pressure in excess of 450 lb would cause the valve to open.

Example 2. An I-beam 20 ft [6.0 m] long and weighing 1600 lb [7.2 kN] rests on rollers placed in the positions shown in Figure 4-12b. Compute the forces exerted on the pavement by the rollers.

Solution: Calling the left roller R_1 and the right R_2, we may write an equation of moments, taking the right roller as the center of moments.

Note that the center of gravity of the beam lies 6 ft [1.8 m] from R_1 and 4 ft [1.2 m] from R_2. Thus

$$R_2 \times 10 = 1600 \times 6 \qquad \left[\begin{array}{c} R_2 \times 3 = 7.2 \times 1.8 \\ R_2 = 4.32 \text{ kN} \end{array}\right.$$
$$R_2 = 960 \text{ lb}$$

Then

$$R_1 + R_2 = 1600 \qquad \left[\begin{array}{c} R_1 + R_2 = 7.2 \\ 4.32 + R_1 = 7.2 \\ R_1 = 2.88 \text{ kN} \end{array}\right]$$
$$960 + R_1 = 1600$$
$$R_1 = 640 \text{ lb}$$

Example 3. Figure 4-12c indicates a capstan in plan. If two men exert cranking forces of 150 lb [675 N] each at the positions shown, what tensile force P will be exerted in the rope?

Solution: Note that in the diagram the dimensions are given in both feet and inches. In writing moment equations be careful to have all the forces and lengths in the same unit. (All pounds and feet; all newtons and meters; etc.) Taking the center of the capstan as the center of moments, we write

$$(150 \times 36) + (150 \times 36) = P \times 10$$
$$P = 1080 \text{ lb}$$
$$\left[\begin{array}{c} (675 \times 900) + (675 \times 900) = P \times 250 \\ P = 4860 \text{ N} \end{array}\right]$$

The two forces of 150 lb each constitute a mechanical couple. In Section 4-3 we found that the moment of a couple is the magnitude of one of the forces multiplied by the perpendicular distance between their lines of action. In this instance the moment of the force P is $(10 \times P)$ in units of inch-pounds. Thus we might have said

$$72 \times 150 = 10 \times P \quad \text{and} \quad P = 1080 \text{ lb}$$
$$[1.8 \times 675 = 250P \quad \text{and} \quad P = 4860 \text{ N}]$$

Example 4. A wheel and axle are indicated in Figure 4-12d. Compute the magnitude of the force P that will balance the force of 200 lb [900 N].

Solution: With the center of the axle as the center of moments, we may write

$$P \times 2 = 200 \times 10 \qquad \left[\begin{array}{c} P \times 50 = 900 \times 250 \\ P = 4500 \text{ N} \end{array} \right.$$

$$P = 1000 \text{ lb}$$

Theoretically, the force of 200 lb will hold the 1000 lb force in equilibrium and will raise any force of a lesser magnitude.

Problem 4-7-A*.   A lever is 5 ft [1.5 m] in length and has a fulcrum (support) 6 in. [0.15 m] to the right from a weight to be lifted. The weight is at the left end of the lever. If the weight is 1000 lb [4.5 kN], what is the magnitude of the downward force at the right and of the lever that will be required to lift the weight?

Problem 4-7-B.   A beam 10 ft [3 m] in length has a concentrated load of 600 lb [2.7 kN] at 2 ft [0.6 m] from one end. Compute the magnitude of the reactions.

Problem 4-7-C.   A beam 16 ft [4.8 m] in length and weighing 3680 lb [16 kN] is suspended by two cables. One cable is 2 ft [0.6 m] from one end of the beam, and the other cable is 4 ft [1.2 m] from the other end. Compute the magnitudes of the forces in the cables.

Problem 4-7-D.   A beam 12 ft [3.6 m] in length and weighing 2400 lb [10 kN] has a vertical concentrated load of 100 lb [0.4 kN] at one end. Compute the distance from the concentrated load at which a single support can be placed to balance the beam and load.

Problem 4-7-E.   Assume that the weight in Figure 4-12a is 32 in. [0.8 m] from the valve. What pressure would be required to open the valve?

Problem 4-7-F.   For the beam shown in Figure 4-12b compute the magnitudes of the reactions if the left roller is moved to the left end of the beam.

Problem 4-7-G*.   For the beam shown in Figure 4-12b compute the magnitudes of the reactions if the right roller is placed at 10 ft [3 m] from the right end of the beam.

Problem 4-7-H.   Figure 4-12c indicates a capstan. What should the magnitude of the two cranking forces be to maintain equilibrium if the force *P* is 1800 lb [8 kN]?

Problem 4-7-I.   For the wheel and axle shown in Figure 4-12d what weight should be applied to raise a 1200-lb [5.3 kN] load at *P* if the diameter of the wheel is 25 in. [625 mm] instead of 20 in.?

Problem 4-7-J*.   A beam is 8 ft [2.4 m] in length and weighs 9600 lb [42 kN]. There are three concentrated loads of 100, 200, and 300 lb [450, 900, and 1350 N] placed at 2, 5, and 8 ft [0.6, 1.5, and 2.4 m], respectively from the left support. What is the distance of the center of gravity of the weight of the beam and the loads from the left end of the beam?

5

Stress and Deformation
IIIIIIIIIIIIIIIIIIIIIIIIIIIIIIIIIIIIIIIIIIIIIIIIII

5-1. Mechanical Properties of Materials

When forces act on a body, internal resisting forces are developed within the body by stresses in the material of the body. Stresses are generally visualized as unit stresses and are measured in terms of force per unit area. The unit area is usually an increment of the area of a cross section of the body, and the force is that required to be developed at the cross section. Thus in Figure 2-1a, the unit stress in the wood piece supporting the external 6400 lb load was found to be 6400/64 = 100 psi; 64 being the area of the cross section of the wood piece in square inches and 6400 being the required internal resisting force at a cross section.

Internal force actions include compression, tension, shear, bending, and torsion. These may occur singly or in various combinations, and many different situations are considered throughout this book. At the moment let us consider only the simple actions of tension and compression, which are called direct forces. The stresses produced by direct forces are called direct stresses (see Figures 2-1 and 2-2). For direct stresses, computations take the form

$$P = f \times A \quad \text{or} \quad f = \frac{P}{A} \quad \text{or} \quad A = \frac{P}{f}$$

in which

P = axial direct load in pounds or newtons
f = unit stress in pounds per square inch or pascals
A = area of the loaded cross section in square inches or square millimeters

For design of a structure, it is necessary to first compute the force or forces that the members of the structure must resist and then to determine the material, size, and shape for each member. A member must be sufficiently large, but waste of material will result if the member is too large for a given task. Size increase will generally lower the unit stresses in the material, and the designer is immediately confronted with the question, "What is the proper allowable unit stress to use in the computations?" Structural materials vary in quality and in basic physical and mechanical properties; hence it behooves the designer to have some knowledge of the nature of commonly used materials. For structural considerations some properties of concern are weight, thermal response, strength, stiffness, elasticity, ductility, hardness, malleability, and brittleness.

5-2. Strength

The *strength* of a material is its ability to resist forces. The three basic stresses are compression, tension, and shear; hence, in speaking of the strength of a material, we must know the type of stress to which the material is to be subjected. As an example, the compressive and tensile strengths of structural steel are about equal, whereas cast iron is strong in compression and relatively weak in tension. The *ultimate strength* of a material is the unit stress that causes failure or rupture. The term *elastic strength* is sometimes applied to the greatest unit stress a material can resist without a permanent change in shape.

5-3. Stiffness

The *stiffness* of a material is the property that enables it to resist deformation. If, for instance, blocks of steel and wood of equal size are subjected to equal compressive loads, the wood block will become shorter than the steel block. The deformation (shortening) of the wood will probably be 20 times that of the steel, and, we say, steel is *stiffer* than wood.

5-4. Elasticity

Elasticity is the property of a material that enables it to return to its original size and shape when the load to which it has been subjected is removed. This property varies greatly in different materials. For certain materials there is a unit stress beyond which the material does not regain its original dimensions when the load is removed. This unit stress is called the *elastic limit*; the allowable (desirable) unit stresses for such materials should be well below the elastic limit. Every material changes its size and shape when subjected to loads. For the materials used in building construction the actual unit stresses should be such that these deformations for direct stresses are in direct proportion to the applied loads.

Plasticity is the opposite quality to elasticity. A perfectly plastic material is a material that does not return to its original dimensions when the load causing deformation is removed. There are probably no perfectly plastic materials. Modeling clay and lead are examples of plastic materials.

5-5. Ductility

Ductility is the property of a material that permits it to undergo plastic deformation when subjected to a tensile force. A material that may be drawn into wires is a ductile material. A chain made of ductile material is preferable to a chain in which the material is brittle.

5-6. Malleability

A material having the property that permits plastic deformation when subjected to a compressive force is a *malleable* material. Materials that may be hammered into sheets are examples of malleable materials. Ductile materials are generally malleable. A material, such as cast iron, that is neither malleable nor ductile is called *brittle*.

5-7. Materials Used in Construction

In a book of this scope it is not possible to present a detailed discussion of the wide range of materials used in building construction. Consequently, we direct our attention mainly to the principal structural materials—steel, concrete, and wood.

Steel varies greatly in its physical properties. These properties are controlled by the method of manufacture and the chemical composition. The percentage of carbon in steel affects both its strength and hardness. In addition to the carbon steels, there are many steel alloys. These alloys are made by the addition of certain other metals, such as nickel and chromium, in various proportions. Steel alloys are employed when certain mechanical properties which are not afforded by the usual carbon steels are desired. Steel meeting the requirements of the American Society for Testing and Materials Specification A36 is the grade of structural steel commonly used for building construction.

The ingredients of concrete are so varied in kind and proportions that concrete of various strengths and properties may be produced. The designer of a reinforced concrete structure bases his computations on the use of concrete having a specified compressive strength (2500, 3000, 3500, etc., psi) at the end of a 28-day curing period. Methods of proportioning concrete mixes to attain desired strengths are specified in recommended practice standards issued by the American Concrete Institute and the Portland Cement Association.

Timber, unlike steel or concrete, is not a processed material but an organic material generally used in its natural state. The many species and grades of wood, the variation in moisture content, and the presence of natural defects (knots, checks, slope of grain, etc.) require special attention in establishing the strength grades of structural lumber. Standards in this connection are promulgated through the National Forest Products Association.

The strength of masonry walls and piers depends not only on the strength of the masonry units such as bricks, stone, and concrete blocks, but also on the kind of mortar that is used.

5-8. Deformation

Whenever a body is subjected to a force, there is a change in the shape or size of the body. These changes in dimensions are called *deforma-*

tions. A block subjected to a compressive force *shortens* in length, and the decrease in length is its deformation. When a tensile force is applied to a rod, the original length of the rod is increased, and the *lengthening* or *elongation* is its deformation. A loaded beam resting on two supports at its ends tends to become concave on its upper surface; we say the beam *bends*. The deformation that accompanies bending is called *deflection*.

When stresses occur in a body, there is always an accompanying deformation. The deformation often is so small that it is not apparent to the naked eye; nevertheless, it is always present. In building construction the deformations in most of the structural members are generally so small that they require no special attention. However, the deflection of beams does demand consideration. A beam should be large enough to resist properly the stresses set up in its fibers, and, what is equally important, the beam should be large enough to prevent the deflection from exceeding certain allowable limits.

The term *strain* is sometimes used as a synonym for deformation. We use the phrase ''stress and strain,'' for which we may substitute ''stress and deformation.''

Consider a wrought iron rod 20 ft 0 in. in length that is subjected to a tensile force. We find that the rod is 0.15 in. longer than it was before the load was applied. The 0.15 in. is the *total elongation*. Since the rod had an original length of 20 ft 0 in., or 240 in., the *unit elongation*, that is, the elongation *per unit of length*, is 0.15 ÷ 240, or 0.00062 in./in.

5-9.   Elastic Limit, Yield Point, and Ultimate Strength

The *elastic limit* is a unit stress that has special significance with respect to certain building materials. The following discussion of a test made on a specimen shows why particular attention must be given to this unit stress.

Laboratories devoted to the testing of materials have machines in which tensile tests are made. Such a machine contains screws operated by motors. When the specimen to be tested is secured in gripping devices, the motors are set in motion, and tensile forces are thus applied to the specimen. The stresses in the specimen are read by means of a counterpoise and scale, and the accompanying elongations of the specimen are read by an instrument called an extensometer. This instrument permits the reading of minute changes in the length of the specimen. During the testing of a specimen a constant record is kept of the applied loads and the resulting deformations.

For the purpose of illustration, let us imagine that a tensile test is made on a rod of structural steel 1 in.^{2} in cross section. Two marks, exactly 8 in. apart, are made on the specimen, and the elongations of this marked length are recorded as the test proceeds. When a load of 5000 psi has been attained, the extensometer reading shows that the total elongation is 0.00139 in. When the next unit of 5000 psi is reached (making a total of 10,000 psi), the total elongation is 2 × 0.00139 in. For an additional 5000-lb load (making a total of 15,000 psi) the total deformation is 3 × 0.00139 in. These stresses and deformations are recorded as shown in Table 5-1. For a unit stress of 35,000 lb the total deformation is 7 × 0.00139 in., or 0.00973 in. The test is continued, and for the next 5000-lb additional load we would expect to see a deformation of 8 × 0.00139 in., or 0.01112 in. However, we find that the deformation for this stress of 40,000 psi is 0.0995 in.

Note that up to and including the unit stress of 35,000 psi the deformations have been directly proportional to the applied loads. The unit stress beyond which the deformations increase in a faster ratio than the applied loads is called the *elastic limit*. The elastic limit is sometimes called the *proportional limit*. For ordinary structural steel it is approximately 36,000 psi [250 MPa].

If during the test, up to the elastic limit, the loads had been removed from the specimen, the bar would have returned to its original length.

TABLE 5-1. Tensile Test

Stress (psi)	Total Deformation (in.)
5,000	0.00139 = 0.00139
10,000	2 × 0.00139 = 0.00278
15,000	3 × 0.00139 = 0.00417
20,000	4 × 0.00139 = 0.00556
25,000	5 × 0.00139 = 0.00695
30,000	6 × 0.00139 = 0.00834
35,000	7 × 0.00139 = 0.00973
40,000	= 0.09950

Sometimes the elastic limit is defined as that unit stress beyond which the stressed member will not return to its original length when the load is removed. If it does not return to its original length, there is a permanent deformation, called a *permanent set*.

In testing certain materials we find that at a stress slightly above the elastic limit a deformation occurs without any increase in stress. The stress at which this deformation occurs is called the *yield point*.

When the stresses and their accompanying deformations during a test have been recorded, it is convenient to plot them on a sheet of graph paper. In Figure 5-1 the vertical distances (ordinates) are the unit stresses, and the horizontal distances (abscissas) show the deformations. Since up to a stress of about 36,000 psi (the elastic limit) the deformations were directly proportional to the stresses, the stress-deformation curve is a straight line. At a stress slightly beyond the elastic limit there is a deformation without an additional stress. This stress, the yield point, is shown graphically by the curve extending a slight distance horizontally.

As increased loads are applied, the curve tends to flatten, and the

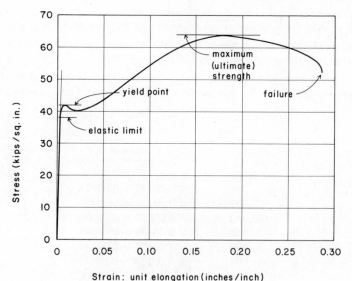

FIGURE 5-1.   Stress-deformation diagram.

maximum stress recorded is the *ultimate strength* of the material. For a steel specimen there is then a decrease in diameter (necking) at one section of a bar and finally rupture or the *breaking strength*. The *ultimate strength* of a material is defined as the unit stress that occurs just at or before rupture. For structural steel, the ultimate strength varies from 58,000 to 80,000 psi [400 to 550 MPa].

All materials do not behave in the manner just described. Cast iron, concrete, and wood, as examples, have no yield points, and their elastic limits are difficult to determine. They do not neck, and their ultimate and breaking strengths are identical.

5-10. Modulus of Elasticity

The discussion of the tensile test on a steel specimen (Section 5-9) states that the elongation for each additional 5000-psi unit is an additional elongation of 0.00139 in. This is an approximation used for the purpose of illustration. There are always slight deviations in the deformations that result from inaccuracies in the instruments as well as from observation. However, accepting these figures, note that 0.00139 in. is the *total deformation*, the deformation for an 8-in. length. Thus the *unit deformation* is 0.00139 ÷ 8, or 0.0001737 in. *per inch of length*.

The *modulus of elasticity* of a material is the ratio of the unit stress to the unit deformation. It is represented by the letter E and is an indication of the *stiffness* of a material. For instance, in the test previously described the unit deformation for a unit stress of 5000 psi is 0.0001737 in. Thus the modulus of elasticity of the material is

$$E = \frac{\text{units stress}}{\text{unit deformation}} = \frac{5000}{0.0001737} = 28,785,262 \text{ psi [200 GPa]}$$

The modulus of elasticity is the unit stress divided by the unit deformation, and since the unit stress is in units of pounds per square inch and the unit deformation is an abstract number (inches divided by inches), the modulus of elasticity is in units of pounds per square inch.

The value of E for steel is taken as 29,000,000 psi and for wood, depending on the species and grade, it varies from something less than 1,000,000 psi to about 1,900,000 psi. The modulus of elasticity of concrete ranges from about 2,000,000 to 5,000,000 psi and over depending on the compressive strength. Let

     E = modulus of elasticity of a material in pounds per square inch
     P = applied force in pounds
     f = unit stress in the member in pounds per square inch
     A = area of cross section of the member in square inches
     l = length of member in inches
     e = total deformation in inches
     s = unit deformation in inches per inch

Then

$$E = \frac{\text{unit stress}}{\text{unit deformation}} = \frac{f}{s} = \frac{P/A}{e/l} = \frac{Pl}{Ae}$$

or

$$e = \frac{Pl}{AE}$$

By the use of this formula, we are able to determine the deformation of a member subjected to direct stresses, provided that we know the modulus of elasticity of the material and that *the unit stress does not exceed the elastic limit of the material.* There are five terms in this equation, and if any four are known the remaining term may be computed. In using this formula, be particular to see that l, the length of the member, is in units of inches.

Example 1.   A steel rod 1 in. in diameter and 10 in. in length elongates 0.0069 in. when subjected to a tensile load of 16,000 lb. Compute the modulus of elasticity.
Solution: The cross-sectional area of a rod whose diameter is 1 in. is 0.7854 in.^{2} Then $f = P/A = 16,000/0.7854 = 20,372$ psi. Since the actual unit stress is within the elastic limit of the material, the formula for finding the modulus of elasticity is applicable. Thus

$$E = \frac{Pl}{Ae} = \frac{16,000 \times 10}{0.7854 \times 0.0069}$$

$$= 29,520,000 \text{ psi, the modulus of elasticity}$$

Example 2.   A 0.5-in. square steel rod 10 ft 0 in. in length is subjected to a tensile load of 4800 lb. Compute the total elongation.

Solution: To begin, let us see whether or not the foregoing formula is valid. The area of a 0.5-in.^{2} rod is 0.5 × 0.5, or 0.25 in.^{2} Then f = P/A = 4800/0.25 = 19,200 psi, the actual unit stress. Since this stress is below the elastic limit of the material, about 36,000 psi, the formula may be employed. The length of the rod is 10 ft 0 in., or 120 in., and E for steel is 29,000,000 psi. Then

$$e = \frac{Pl}{AE} = \frac{4800 \times 120}{0.25 \times 29,000,000} = 0.0795 \text{ in., the total elongation}$$

16 ft, 119 in

Problem 5-10-A.   A steel bar 0.5 in. [13 mm] in diameter and 8 in. [200 mm] long elongates 0.0042 in. [0.107 mm] under a tensile load of 3000 lb [13.35 kN]. Compute the modulus of elasticity.

Problem 5-10-B*.   A steel bar has a length of 16 ft [4.8 m] and a cross-sectional area of 0.5 × 1 in. [13 × 25 mm]. What will be its total length under a tensile load of 9000 lb [40 kN]?

Problem 5-10-C.   A timber block 2 × 2 in. [50 × 50 mm] in cross section and 6 in. [150 mm] in length is subjected to a compressive force of 5000 lb [22 kN]. The block shortens 0.005 in. [0.125 mm]. Compute the modulus of elasticity for the material.

Problem 5-10-D*.   If the bar of Problem 5-10-B is made of aluminum alloy having a modulus of elasticity of 10,000,000 psi [70 GPa], calculate its length under load.

5-11.   Factor of Safety

The degree of uncertainty that exists, both with respect to actual loading of a structure and uniformity in the quality of materials, requires that some reserve strength be provided. The degree of reserve strength is known as the *factor of safety*. Although there is no general agreement on the precise definition of this term, the following discussion serves to fix the concept in mind.

Consider a structural steel that has an ultimate tensile strength of 70,000 psi, a yield point stress of 36,000 psi, and an *allowable unit stress* in tension of 22,000 psi. If the factor of safety is defined as the ratio between the ultimate strength and the allowable stress, its value is 70,000 ÷ 22,000 or 3.18. On the other hand, if it is defined as the ratio of the yield point to the allowable stress, its value is 36,000 ÷ 22,000 or 1.64. This is a considerable variation, and since failure of structural member begins when it is stressed beyond the elastic limit, the higher value may be misleading. Consequently, the term *factor of safety* is not employed extensively today.

5-12.   Allowable Unit Stresses

An *allowable unit stress* is the maximum unit stress considered desirable in a structural member subjected to loads. The term *working stress* is also employed to denote the permissible value of a unit stress for use in structural design. The procedures for establishing allowable unit stresses in tension, compression, shear, and bending are different for different materials and are prescribed in specifications promulgated by the American Society for Testing and Materials. In general, allowable unit stresses for structural steel are expressed as fractions of the yield stress, those for wood involve an adjustment of clear wood strength as modified by lumber grading rules and conditions of use, and allowable stresses for concrete are given as fractions of the specified compressive strength of the concrete.

Tables 5-2, 5-3, and 17-1 give allowable stresses for steel, wood, and reinforced concrete construction, respectively, as recommended by the industry associations concerned. When scanning these tables, note that they contain several terms that have not been introduced in this book thus far; these are discussed in subsequent articles as need for the data arises.

In actual structural design practice, the building code governing the construction of buildings in the particular locality must be consulted for specific requirements. Codes are not uniform in their requirements, and the building regulations of one city may be quite different from those of another city nearby. Also, many municipal building codes are revised infrequently and consequently may not be in agreement with current editions of the industry-recommended allowable stresses. Unless otherwise noted, the allowable stresses used in this book are those given in the tables just referenced.

5-13.   Application of Allowable Unit Stresses

The following examples illustrate the application of allowable stresses to the solution of some typical structural problems.

Example 1.   A 1-in. [25 mm] diameter steel rod is used as a hanger to support part of a balcony in an auditorium. If the rod is made of A36 steel, find the maximum allowable load it can carry, assuming stress on the gross area of the rod to be critical. (*Note*: the condition of stress at net cross sections through holes, at threads, and at other reduced cross sections is discussed in Section 14-4.)

Solution: Since the tension in the rod is a direct stress (Section 5-1), the expression $P = f \times A$ applies. Referring to Table 5-2, the allowable unit stress in tension is given as 22 ksi [150 MPa]. The area of the rod cross section is $\pi D^2/4 = 3.1416 \times (1)^2/4 = 0.7854$ in.^{2} [491 mm^2]. Then

$$P = f \times A = 22 \times 0.7854 = 17.28 \text{ k [73,65 kN]}$$

Example 2.   It is proposed to suspend a load of 12 kips [53 kN] from a hanger consisting of a $\frac{7}{8}$-in. [22 mm] diameter rod of A36 steel. Will the gross section of the rod be overstressed under this loading?

Solution: For problems of this kind the procedure is to compute the actual unit stress developed under the loading and compare it with the allowable. The area of the rod is

$$A = \pi D^2/4 = 3.1416 (0.875)^2/4 = 0.6013 \text{ in.}^2 \text{ [380 mm}^2\text{]}$$

Then, using the direct stress relationship (Section 5-1)

$$f = \frac{P}{A} = \frac{12}{0.6013} = 19.96 \text{ ksi [139 MPa]}$$

As this is less than the allowable stress of 22 ksi [150 MPa] noted in the previous example, the rod is not overstressed.

Example 3.   A 10 × 10-in. short post of Douglas Fir Select Structural grade is used to support a compressive load. Compute the value of the allowable load, assuming that the direction of the load is parallel to the grain of the wood.

Solution: We note that the actual dimensions of the post are 9.5 × 9.5 in. [240 × 240 mm] and the area of the cross section is 90.25 in.^{2} [57,600 mm^2] (Table 6-7). Referring to Table 5-3 under Douglas Fir-Larch, we locate the size classification of "posts and timbers" and the line for Select Structural grade and note the allowable stress for compression parallel to the grain (given in the eighth column) is 1150 psi [7.93 MPa]. Then

$$P = f \times A = 1150 \times 90.25 = 103.788 \text{ lb [457 kN]}$$

Problem 5-13-A.   Will a round rod of A36 steel 1.25 in. [32 mm] in diameter safely support a tensile load of 24 k [107 kN]?

TABLE 5-2. Allowable Unit Stresses for Structural Steel: ASTM A36[a]

Type of Stress and Conditions	Stress Designation	AISC Specification	Allowable Stress (ksi)	Allowable Stress (MPa)
Tension				
1. On the gross (unreduced) area	F_t	$0.60 F_y$	22	150
2. On the effective net area, except at pinholes		$0.50 F_u$	29	125
3. Threaded rods on net area at thread		$0.33 F_u$	19	80
Compression	F_a	See discussion in Chapter 13		
Shear				
1. Except at reduced sections	F_v	$0.40 F_y$	14.5	100
2. At reduced sections		$0.30 F_u$	17.4	120
Bending				
1. Tension and compression on extreme fibers of compact members braced laterally, symmetrical about and loaded in the plane of their minor axis	F_b	$0.66 F_y$	24	165

2. Tension and compression on extreme fibers of other rolled shapes braced laterally	$0.60 F_y$	22	150
3. Tension and compression on extreme fibers of solid round and square bars, on solid rectangular sections bent on their weak axis, on qualified doubly symmetrical I & H shapes bent about their minor axis	$0.75 F_y$	27	188
Bearing	F_p		
1. On contact area of milled surfaces	$0.90 F_y$	32.4	225
2. On projected area of bolts and rivets in shear connections	$1.50 F_u$	87	600

[a] F_y = 36 ksi; assume that F_u = 58 ksi; some table values are rounded off as permitted in the AISC Manual. For SI units F_y = 250 MPa. F_u = 400 MPa.

TABLE 5-3. Allowable Unit Stresses for Structural Lumber—Visual Grading[a]

Species and Commercial Grade	Size Classification	Extreme Fiber in Bending F_b		Tension Parallel to Grain F_t	Horizontal Shear F_v	Compression Perpendicular to Grain F_{c1}	Compression Parallel to Grain F_c	Modulus of Elasticity E
		Single-Member Uses	Repetitive-Member Uses					
Douglas Fir–Larch (Surfaced dry or surfaced green. Used at 19% max moisture content)								
Dense Select Structural	2–4 in. thick, 5 in. and wider	2100	2400	1400	95	730	1650	1,900,000
Select Structural		1800	2050	1200	95	625	1400	1,800,000
Dense No. 1		1800	2050	1200	95	730	1450	1,900,000
No. 1		1500	1750	1000	95	625	1250	1,800,000
Dense No. 2		1450	1700	775	95	730	1250	1,700,000
No. 2		1250	1450	650	95	625	1050	1,700,000
No. 3		725	850	375	95	625	675	1,500,000
Dense Select Structural	beams and stringers	1900		1100	85	730	1300	1,700,000
Select Structural		1600		950	85	625	1100	1,600,000
Dense No. 1		1550		775	85	730	1100	1,700,000
No. 1		1300		675	85	625	925	1,600,000
Dense Select Structural	posts and timbers	1750		1150	85	730	1350	1,700,000
Select Structural		1500		1000	85	625	1150	1,600,000
Dense No. 1		1400		950	85	730	1200	1,700,000
No. 1		1200		825	85	625	1000	1,600,000

Southern Pine (Surfaced dry. Used at 19% max moisture content)

Select Structural	2–4 in. thick, 5 in. and wider	1750	2000	1150	90	565	1350	1,700,000
Dense Select Structural		2050	2350	1300	90	660	1600	1,800,000
No. 1		1450	1700	975	90	565	1250	1,700,000
No. 1 Dense		1700	2000	1150	90	660	1450	1,800,000
No. 2		1200	1400	625	90	565	1000	1,600,000
No. 2 Dense		1400	1650	725	90	660	1200	1,600,000
No. 3		700	800	350	90	565	625	1,400,000
No. 3 Dense		825	925	425	90	660	725	1,500,000

Southern Pine (Surfaced green. Used in any condition)

Dense Structural 86	2.5 in. and thicker	2100	2400	1400	145	440	1300	1,600,000
Dense Structural 72		1750	2050	1200	120	440	1100	1,600,000
Dense Structural 65		1600	1800	1050	110	440	1000	1,600,000
No. 1 SR	5 in. and thicker	1350		875	110	375	775	1,500,000
No. 1 Dense SR		1550		1050	110	440	925	1,600,000
No. 2 SR		1000		725	95	375	625	1,400,000
No. 2 Dense SR		1250		850	95	440	725	1,400,000

[a] Stresses listed are for normal loading conditions.

Source: Adapted from more extensive tables in *Design Values for Wood Construction*, 1982 ed., with permission of the publishers, National Forest Products Association.

Problem 5-13-B.   A bar of A36 steel 1 × 3 in. [25 × 75 mm] is used as a hanger. What total load may be supported?

Problem 5-13-C*.   Will a short post of Douglas Fir No. 1 grade support a load of 20 kips [89 kN] if the post nominal size is 6 × 6 in.?

Problem 5-13-D.   A short post of Douglas Fir Dense No. 1 grade has nominal dimensions of 8 × 8 in. Compute the maximum allowable compressive load the post will support.

6

Properties of Sections
||||||||||||||||||||||||||||||||||||||||||||||

6-1. Centroids

The *center of gravity* of a solid is an imaginary point at which all its
weight may be considered to be concentrated or the point through which
the resultant weight passes. Since an area has no weight, it has no
center of gravity. *The point of a plane area that corresponds to the
center of gravity of a very thin homogeneous plate of the same area
and shape is called the centroid of the area.*

When a simple beam is subjected to forces that cause it to bend, the
fibers above a certain plane in the beam are in compression and those
below the plane are in tension. This plane is called the *neutral surface*.
For a cross section of the beam the line corresponding to the neutral
surface is called the *neutral axis*. The neutral axis passes through the
centroid of the section; thus it is important that we know the exact
position of the centroid.

The position of the centroid for symmetrical shapes is readily determined. If an area possesses a line of symmetry, the centroid will
obviously be on that line; if there are two lines of symmetry, the centroid will be at their point of intersection. For instance, a rectangular
area as shown in Figure 6-1a has its centroid at its geometrical center,
the point of intersection of the diagonals. The centroid of a circular
area is its center (Figure 6-1b). With respect to a triangular area (Figures 6-1c and d), it is convenient to remember that the centroid is at a

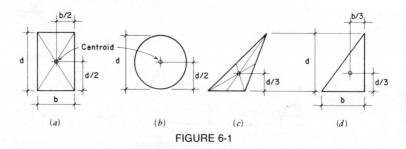

FIGURE 6-1

distance equal to one-third of the perpendicular distance measured from any side to the opposite vertex. The intersection of the lines drawn from the vertices to the midpoints of the opposite sides is another method of locating the centroid of a triangle. This is shown in Figure 6-1c.

Note: Tables 6-1 through 6-7, referred to in the following discussions, are presented at the end of this chapter.

For symmetrical structural steel shapes, such as I-beams, the centroid is on the vertical axis through the web at a point midway between the upper and lower surfaces of the flanges. A 12-in. standard I-beam has the centroid of its cross section 6 in. from the top or bottom of the section, at the intersection of the X–X and Y–Y axes shown in the diagram of Table 6-2. Similar tables of properties are available for unsymmetrical structural steel sections. Table 6-3, for example, gives the properties of channel sections. Consider a 10-in. channel having a weight of 15.3 lb per lin ft (designated C 10 × 15.3). In this table we find that the vertical axis is 0.634 in. from the back of the web, the distance $\bar{x}$ (read x bar). The centroid is on this axis at a point 5 in. between the top and bottom surfaces of the flanges. The position of the centroid of angle sections is given in Tables 6-4 and 6-5.

It frequently happens, however, that we must determine the position of the centroid, and this is accomplished most readily by mathematics. The *statical moment* of a plane area with respect to a given axis is the area multiplied by the normal distance of the centroid of the area to the axis. *If an area is divided into a number of parts, the sum of the statical moments of the parts is equal to the statical moment of the entire area.* This is the principle by means of which the position of the centroid is found; its application is remarkably simple.

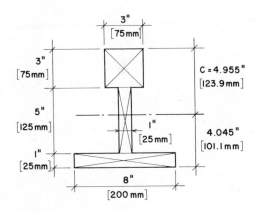

$$M_x = Area * C(y)$$

$$M_y = Area * c(x)$$

FIGURE 6-2

Example. Figure 6-2 is a beam cross section, unsymmetrical with respect to the horizontal axis. Find the value of c, the distance of the neutral axis from the most remote fiber.

Solution: (1) It is not always possible to tell by observation whether the centroid is nearer the top or bottom edge of the area. In this instance let us write an equation of moments about an axis through the uppermost edge. First divide the total area into some number of simple shapes, in this case the three rectangles shown by the diagonals. The area of the upper part is 9 in.^{2} [5625 mm^2], and its centroid is 1.5 in. [37.5 mm] from the reference axis. The center part has an area of 5 in.^{2} [3125 mm^2], and its centroid is 5.5 in. [137.5 mm] from the axis. The bottom part contains 8 in.^{2} [5000 mm^2], and its centroid is 8.5 in. [212.5 mm] from the axis. Considering these individual areas as forces, their combined moment about the reference axis is

$$M = (9 \times 1.5) + (5 \times 5.5) + (8 \times 8.5)$$
$$= 13.5 + 27.5 + 68 = 109 \text{ in.}^3$$

Since this is the moment of the entire area about the axis, it may be equated to the product of the entire area (22 in.^{2}) times the single centroid distance c, as shown in the figure. Thus

$$M = 109 = 22 \times c$$
$$c = 109 \div 22 = 4.955 \text{ in. } [123.9 \text{ mm}]$$

which is the distance of the centroid from the reference axis.

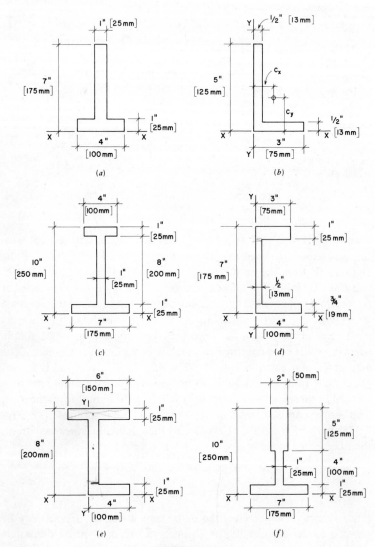

FIGURE 6-3

94

(2) The depth of the section is 9 in [225 mm]; hence the distance of the centroid from the bottom edge is

$$9 - 4.955 = 4.045 \text{ in. } [101.1 \text{ mm}]$$

Call this distance c_1. The value of c_1 may be checked by writing an equation of moments about a reference axis through the bottom edge; thus

$$(9 \times 7.5) + (5 \times 3.5) + (8 \times 0.5) = (22 \times c_1)$$

from which

$$c_1 = \frac{89}{22} = 4.045 \text{ in.}$$

Remember that the centroid of the section is the point through which the neutral axis (zero bending stress point) passes. The position of the centroid for structural steel angles may be found directly from Table 6-4. The locating dimensions from the backs of the angle legs are given as y and x distances.

Problems 6-1A*-B-C-D*-E-F.   Find the location of the centroid for the cross-sectional areas shown in Figure 6-3. Use the references and indicate the distances as c_x and c_y, as shown in Figure 6-3b.

6-2. Moment of Inertia

Consider the area enclosed by the irregular line in Figure 6-4a, area A. In this area an infinitely small area a is indicated at z distance from an axis marked $X–X$. If we multiply this tiny area by the square of its distance to the axis, we have the quantity $(a \times z^2)$. The entire area is made up of an infinite number of these small elementary areas at various distances from the $X–X$ axis. Now, if we use the Greek letter Σ to represent the sum of an infinite number, we may write Σaz^2, which indicates the sum of all the infinitely small areas (of which the area A is composed) multiplied by the square of their distances from the $X–X$ axis. This quantity is called the *moment of inertia* of the area and is represented by the letter I. Then $I_{X–X} = \Sigma az^2$, meaning that Σaz^2 is the moment of inertia of the area with respect to the axis marked $X–X$.

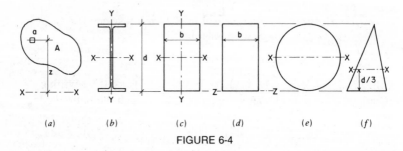

FIGURE 6-4

We may define the moment of inertia of an area as *the sum of the products of all the elementary areas multiplied by the square of their distances from an axis*. In the United States the linear dimensions of cross sections of structural members are invariably in units of inches, and, since the moment of inertia involves an area multiplied by the square of a distance, the moment of inertia is expressed in inches to the fourth power and is written in.[4], sometimes called biquadratic inches. We cannot conceive such units as, for example, a linear dimension; it is a mathematical concept, expressed as Σaz^2. Thus the moment of inertia of a cross section depends not only on the number of units in its area but also on the arrangement of the area with respect to the axis under consideration. The moment of inertia of cross sections is of particular importance in the design of beams and columns. It is important to realize that the moment of inertia of cross sections of beams is taken about an axis that passes through the centroid of the area; very often such a moment of inertia is represented as I_0. Figure 6-4b shows the section of a 12-in. standard I-beam weighing 31.8 lb/lin ft (designated S 12 × 31.8). The horizontal axis is *X–X* and the vertical axis is *Y–Y*. Referring to Table 6-2, we find that $I_{X-X} = 218$ in.[4] and $I_{Y-Y} = 9.36$ in.[4] Obviously, the farther the component parts of an area are from the axis, the greater is the value of the moment of inertia.

The moment of inertia of rolled steel sections used in building construction is given in tables of properties published by the American Institute of Steel Construction. Tables 6-1 through 6-5, given at the end of this chapter, have been compiled from more extensive tables in the AISC *Manual of Steel Construction*.

6-3.   Moment of Inertia of Geometric Figures

It is frequently necessary to compute the moment of inertia for built-up sections or for sections not given in tables. In engineering problems the shapes most commonly involved are the rectangle, the circle, and the triangle. The derivations of the formulas to be used for computing moments of inertia are most readily accomplished by the use of calculus. It is beyond the scope of this book to derive such formulas, but their values are given here and the computations involved in computing I for the sections most frequently used present no difficulties.

Rectangles.   Consider the rectangle shown in Figure 6-4c. Its width is b and its depth is d. The two major axes are X–X and Y–Y, both passing through the centroid of the area. It can be shown that *the moment of inertia of a rectangular area about an axis passing through the centroid parallel to the base is* $I_{X-X} = bd^3/12$. With respect to the vertical axis, $I_{Y-Y} = db^3/12$. This value for the moment of inertia of a rectangle is developed in Section 11-4.

Example 1.   Find the value of the moment of inertia for a 6×12-in. wood beam about an axis through its centroid and parallel to the base of the section.

Solution: Referring to Table 6-7, we find that a nominal 6×12-in. section has standard dressed dimensions of 5.5×11.5 in. [139.7 $\times$ 292.1 mm]. Then

$$I = \frac{bd^3}{12} = \frac{(5.5)(11.5)^3}{12} = 697.07 \text{ in.}^4 \ [290.1 \times 10^6 \text{ mm}^4]$$

which is in agreement with the value of I listed in the table.

The moment of inertia of a rectangular area with respect to an axis of reference passing through its base may be found as $bd^3/3$. This is sometimes useful in finding the moment of inertia of complex sections, as is demonstrated in the following example.

Example 2.   Find the moment of inertia of a $6 \times 4 \times \frac{1}{2}$-in. angle [152.4 $\times$ 101.6 $\times$ 12.7 mm] (Figure 6-5) about its neutral axis, which has been determined to be 2 in. above its base (the back of the 4-in. leg).

Solution: (1) We first divide the section into three parts. The first part

98   Properties of Sections

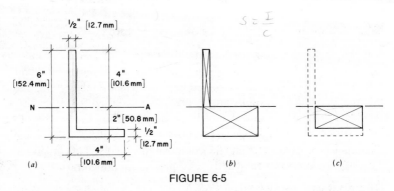

$$S = \frac{I}{c}$$

FIGURE 6-5

is the portion of the 6-in. leg above the neutral axis. The second part is assumed to be a solid rectangle of 2×4 in. as shown in Figure 6-5*b*. The third part is a negative area, which is what must be subtracted from the second part; this area is shown in Figure 6-5*c*. The sum of these parts will produce the net form of the angle.

(2) We now proceed to find the total moment of inertia for the angle by finding the individual moments of inertia of these three parts about their edges. For the first part (see Figure 6.10):

$$I = \frac{bd^3}{3} = \frac{(0.5)(4)^3}{3} = 10.667 \text{ in.}^4 \ [4.440 \times 10^6 \text{ mm}^4]$$

For the second part:

$$I = \frac{(4)(2)^3}{3} = 10.667 \text{ in.}^4 \ [4.440 \times 10^6 \text{ mm}^4]$$

For the third part:

$$I = -\frac{(3.5)(1.5)^3}{3} = 3.938 \text{ in.}^4 \ [1.639 \times 10^6 \text{ mm}^4]$$

And for the entire angle

$$I = 10.667 + 10.667 - 3.938 = 17.396 \text{ in.}^4 \ [7.241 \times 10^6 \text{ mm}^4]$$

which may be compared to the value listed for I_x in Table 6-4.

The preceding example illustrates one method of finding the moment of inertia of an unsymmetrical section, but it is not necessary to compute I for the commonly used structural steel angles, since these may be found directly from Table 6-4 and from the more extensive tables in the AISC Manual.

Circles.   Figure 6-4e shows a circular area with diameter d, and axis X–X passing through its center. *The moment of inertia of a circular area about an axis passing through its centroid is* $I_{X-X} = \pi d^4/64$.

Example 3.   Compute the moment of inertia of a circular cross section, 10 in. in diameter, about an axis through its centroid.
Solution: Since I_{X-X} in Figure 6-4e may be any axis through the centroid of the circle, we may use the symbol I_0. Then

$$I_0 = \frac{\pi d^4}{64} = \frac{3.1416 \times 10^4}{64} = 490.9 \text{ in.}^4 \ [204.3 \times 10^6 \text{ mm}^4]$$

Triangles.   The triangle in Figure 6-4f has a height d and a base b. The axis X–X passes through the centroid of the area. *The moment of inertia of a triangular cross section about an axis through the centroid parallel to the base is* $I_{X-X} = bd^3/36$.

Example 4.   Assuming that the base of the triangular area shown in Figure 6.4f is 12 in. and the height 10 in., compute the moment of inertia about axis X–X which passes through the centroid.
Solution: For this triangle $b = 12$ in. and $d = 10$ in. Then

$$I_0 = \frac{bd^3}{36} = \frac{12 \times 10^3}{36} = 333.33 \text{ in.}^4 \ [138.7 \times 10^6 \text{ mm}^4]$$

Hollow Sections and I-Shapes   The moment of inertia of some geometric figures may be found by employing the principle that states *if the moments of inertia of the component parts of an area are taken about the same axis, they may be added or subtracted to find the moment of inertia of the area.*

Example 5.   Compute the moment of inertia of the hollow rectangular section shown in Figure 6-6a about a horizontal axis through the centroid parallel to the 6-in. side.
Solution: The moment of inertia of the 6×10-in. rectangle is

$$I = \frac{bd^3}{12} = \frac{6 \times 10^3}{12} = 500 \text{ in.}^4 \ [208.11 \times 10^6 \text{ mm}^4]$$

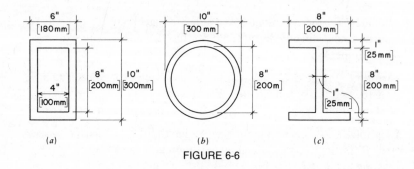

FIGURE 6-6

and that of the 4 × 8-in. rectangle is

$$I = \frac{bd^3}{12} = \frac{4 \times 8^3}{12} = 170.7 \text{ in.}^4 \ [71.03 \times 10^6 \text{ mm}^4]$$

Since both of these moments of inertia are taken *about the same axis*, we may subtract that of the smaller rectangle from that of the larger, and the difference will be the moment of inertia of the hollow rectangular section. Thus $500 - 170.7 = 329.3$ in.^{4} [137.08 × 10^6 mm^4]

Example 6. Compute the moment of inertia about an axis through the centroid of the pipe cross section shown in Figure 6-6b. The thickness of the pipe shell is 1 in.
Solution: To find I_0 of the ring section, we will compute I for the area bounded by the outer circle and then subtract from this value the I for the inner circular area. Using the 10-in. outside diameter of the pipe,

$$I = \frac{\pi d^4}{64} = \frac{3.1416 \times 10^4}{64} = 491 \text{ in.}^4 \ [204.3 \times 10^6 \text{ mm}^4]$$

and working with the 8-in. inside diameter,

$$I = \frac{\pi d^4}{64} = \frac{3.1416 \times 8^4}{64} = 201 \text{ in.}^4 \ [83.7 \times 10^6 \text{ mm}^4]$$

Therefore the moment of inertia of the pipe section is

$$I_0 = 491 - 201 = 290 \text{ in.}^4 \ [120.6 \times 10^6 \text{ mm}^4]$$

Example 7.   Referring to Figure 6-6c, compute the moment of inertia

$$\frac{bd^3}{12} \quad \text{or} \quad \frac{bd^3}{3}$$

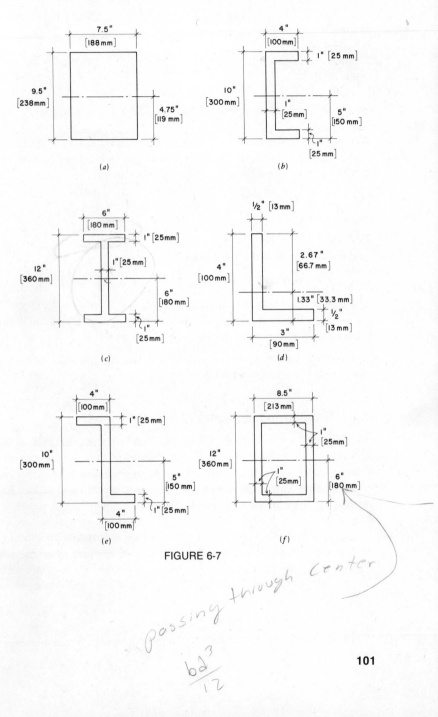

FIGURE 6-7

Passing through Center

$$\frac{bd^3}{12}$$

101

of the I-section about a horizontal axis through the centroid of the section parallel to the flange.

Solution: For the full rectangular area with the width of 8 in. and depth of 10 in.,

$$I = \frac{bd^3}{12} = \frac{8 \times 10^3}{12} = 667 \text{ in.}^4 \ [277.5 \times 10.6 \text{ mm}^4]$$

This moment of inertia includes that of the two open spaces at the sides of the vertical web. Taken together, these open spaces are equivalent to a rectangle 7 in. wide and 8 in. deep. For this area

$$I = \frac{bd^3}{12} = \frac{7 \times 8^3}{12} = 299 \text{ in.}^4 \ [124.3 \times 10.6 \text{ mm}^4]$$

Subtracting I for the open spaces from I for the large rectangle, the moment of inertia of the I-section is

$$I = 667 - 299 = 368 \text{ in.}^4 \ [153.2 \times 10^6 \text{ mm}^4]$$

Formulas for moment of inertia of rectangles, circles, and triangles are given in Figure 6-10, together with formulas for other properties considered in Sections 6-5 and 6-6.

Problems 6-3-A-B*-C-D-E-F*. Compute the moment of inertia about the neutral axes of the beam sections shown in Fig. 6-7.

6-4.   Transferring Moments of Inertia

When shapes are combined to form built-up structural sections, it is necessary to determine the moment of inertia of the built-up section about its neutral axis. This requires transferring the moments of inertia of some of the individual parts from one axis to another and is accomplished by means of the transfer-of-axis equation, which may be stated as follows:

The moment of inertia of a cross section about any axis parallel to an axis through its own centroid is equal to the moment of inertia of the cross section about its own centroidal axis, plus its area times the square of the distance between the two axes. Expressed mathematically,

$$I = I_0 + Az^2$$

In this formula,

I = moment of inertia of the cross section about the required axis

I_0 = moment of inertia of the cross section about its own centroidal axis parallel to the required axis

A = area of the cross section

z = distance between the two parallel axes

These relationships are indicated in Figure 6-8a, where X–X is the centroidal axis of the angle (passing through its centroid) and Y–Y is the axis about which the moment of inertia is to be found.

To illustrate the use of the equation, we may prove that the value of I for a rectangle about an axis through its base is $bd^3/3$. Since I for a rectangle about its centroidal axis is known to be $bd^3/12$, and z in this instance is $d/2$, we may write

$$I = I_0 + Az^2$$

$$I = \frac{bd^3}{12} + \left[bd \times \left(\frac{d}{2}\right)^2 \right]$$

$$I = \frac{bd^3}{12} + \frac{bd^3}{4} = \frac{bd^3}{3}$$

The application of the transfer formula to the steel built-up section shown in Figure 6-8b is illustrated in the following example.

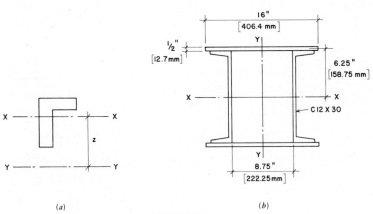

(a)   (b)

FIGURE 6-8

Example. Compute the moment of inertia about the X–X axis of a built-up section composed of two C 12 × 30 channels and two 16 × 0.5-in. [406.4 × 12.7-mm] plates. (See Figure 6-8*b*.)

Solution: (1) From Table 6-3 we find that I_x for the channel is 162 in⁴ [67.42 × 10⁶ mm⁴], so the value for the two channels is twice this, or 324 in.⁴ [134.84 × 10⁶ mm⁴].

(2) For one plate the moment of inertia about an axis through its own centroid is

$$I_0 = \frac{bd^3}{12} = \frac{16 \times 0.5^3}{12} = 0.1667 \text{ in.}^4 \; [0.0694 \times 10^6 \text{ mm}^4]$$

the distance between the centroid of the plate and the X–X axis is 6.25 in. [158.75 mm], and the area of one plate is 8 in.² [5161.3 mm²]. Therefore the *I* of one plate about the X–X axis of the combined section is

$$I_0 + Az^2 = 0.1667 + (8)(6.25)^2 = 312.7 \text{ in.}^4$$

$$[0.0694 \times 10^6 + (5161.3)(158.75)^2 = 130.14 \times 10^6 \text{ mm}^4]$$

and the value for two plates is twice this, or 625 in.⁴ [260.3 × 10⁶ mm⁴].

(3) Adding the moments of inertia of the channels and plates, we obtain the *I* for the entire cross section as 324 + 625 = 949 in.⁴ [(134.84 + 260.3) × 10⁶ = 395.14 × 10⁶ mm⁴].

Problems 6-4-A*-B*-C.   Compute *I* with respect to the X–X axes for the built-up sections shown in Figure 6.9. Make use of any appropriate data given in Tables 6-1 through 6-7.

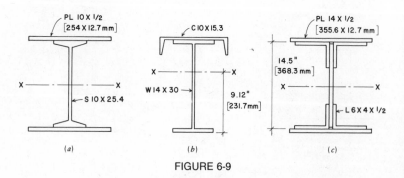

FIGURE 6-9

6-5. Section Modulus

As noted in Section 11-3, the term I/c in the flexure formula is called the *section modulus*. It is defined as the moment of inertia divided by the distance of the most remote fiber from the neutral axis and is denoted by the symbol S. Since I and C always have the same values for any given cross section, values of S may be computed and tabulated for structural shapes. With I expressed in inches to the fourth power and c a linear dimension in inches, S is in units of inches to the third power, written $in.^3$ Section moduli are among the properties tabulated for structural steel shapes in Tables 6-1 through 6-6 and for structural lumber cross sections in Table 6-7.

For a rectangular cross section of breadth b and depth d we know that the moment of inertia about the X–X axis is $bd^3/12$ and that $c = d/2$. Therefore

$$\frac{I}{c} \text{ or } S = \frac{bd^3/12}{d/2} = \frac{bd^3}{12} \times \frac{2}{d}$$

or $S = bd^2/6$. It is often convenient to use this formula directly. It applies, of course, only to rectangular cross sections.

Example 1.   Verify the tabulated value of the section modulus of a 6 × 12-in. wood beam.
Solution: From Table 6-7 we find that the true dimensions of the beam are 5.5 in. × 11.5 in. [139.7 × 292.1 mm]. Then

$$S = \frac{bd^2}{6} = \frac{5.5 \times 11.5 \times 11.5}{6} = 121.23 \text{ in.}^3 \ [1987 \times 10^3 \text{ mm}^3]$$

Compare this with the value of S in Table 6-7.

Example 2.   Verify the tabulated value of S_x for a W 18 × 46.
Solution: From Table 6-1 we find that $I_x = 712$ in.^{4} [296.33 × 10^6 mm^4] and the actual depth is 18.06 in. [458.72 mm]. For the symmetrical section, $c = d \div 2 = 9.03$ in. Then

$$S = \frac{I}{c} = \frac{712}{9.03} = 78.848 \text{ in.}^3 \left[\frac{296.33 \times 10^6}{229.36} = 1292 \times 10^3 \text{ mm}^3 \right]$$

which checks reasonably with the value in the table.

Problems 6-5-A-B-C-D.   Verify the tabulated section modulus values for the following elements.

A. A 6 × 8-in. timber beam. Actually 5.5 × 7.5 in. [139.7 × 190.5 mm].
B. S_x for an S 12 × 31.8 rolled shape.
C. S_x for an L 5 × 3.5 × 0.5 rolled steel angle.
D. S_y for an L 4 × 4 × 0.5 steel angle.

6-6. Radius of Gyration

This property of a cross section is related to the design of compression members rather than beams and is discussed in more detail under the design of columns in subsequent sections of the book. Radius of gyration is considered briefly here, however, because it is a property listed in Tables 6-1 through 6-6.

Just as the section modulus is a measure of the resistance of a beam section to bending, the radius of gyration (which is also related to the size and shape of the cross section) is an index of the stiffness of a structural section when used as a column or other compression member. The radius of gyration is found from the formula

$$r = \sqrt{\frac{I}{A}}$$

and is expressed in inches, since the moment of inertia is in inches to the fourth power and the cross-sectional area is in square inches.

If a section is symmetrical about both major axes, the moment of inertia, and consequently the radius of gyration, is the same for each axis. But most common sections, particularly steel columns, are not symmetrical about the two major axes, and *in the design of columns the least moment of inertia, and therefore the least radius of gyration, is the one used in computations*. By *least* we mean the smallest in magnitude. Note in Table 6-4 that the least radius of gyration of angle sections occurs about the *Z–Z axes*.

Example.   Verify the tabulated values for radii of gyration for a *W* 12 × 58, as given in Table 6-1.
Solution: The table shows the area of this section to be 17.0 in^2 [10968 mm^2] and *I* with respect to the *X–X* axis to be 475 in.^{4} [197.7 × 10^6

mm^4]. Then

$$r_x = \sqrt{\frac{I}{A}} = \sqrt{\frac{475}{17}} = \sqrt{27.94} = 5.29 \text{ in. } [134.3 \text{ mm}]$$

The table value for I with respect to the Y–Y axis is 107 in.^{4} [44.53 × 10^6 mm^4]. Therefore

$$r_y = \sqrt{\frac{I}{A}} = \sqrt{\frac{107}{17}} = \sqrt{6.29} = 2.51 \text{ in. } [63.7 \text{ mm}]$$

Compare these with the values given in Table 6-1. It may be noted that there is a minor discrepancy in the value for r_x, having to do with the manner of rounding off for the last digit.

Problem 6-6-A.   Verify the value of r_y for an S 12 × 40.8.

Problem 6-6-B.   Verify the value of r_x for an L 4 × 3 × $\frac{5}{8}$.

Problem 6-6-C *.   Compute the radius of gyration with respect to the X–X axis for the built-up sections shown in Figure 6-9*b*.

6-7.   Tables of Properties

The preceding sections in this chapter have pertained to the various properties of sections used in the design of structural members. Since the rectangle, circle, and triangle are the areas most commonly involved in engineering problems, the properties of these shapes are given in Figure 6-10. This figure will serve as a reference. Although the properties of commonly used structural sections may be found directly in tables, a thorough knowledge of their significance will be of great value to the designer. Thus far nothing has been said concerning the use of these properties or their application in the design of members that resist forces. This is explained later, particularly in connection with members in bending and in the design of columns. It should be noted here, however, that all of the properties discussed relate to the size and shape of the sections considered—they are independent of the material of which the structural member is made.

For use in design, the properties of selected structural steel shapes are presented in Tables 6-1 through 6-6. Attention is called to the two

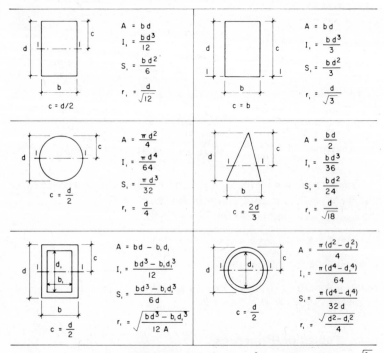

A = Area I = Moment of inertia S = Section modulus = $\frac{I}{c}$ r = Radius of gyration = $\sqrt{\frac{I}{A}}$

FIGURE 6-10.   Properties of common geometric shapes.

major axes, X–X and Y–Y. I, S, and r are given for both axes; the values to be used in design depend on the position in which the member is placed. If an I-section is used as a beam, the web is placed in a vertical position, and therefore the X–X or horizontal axis determines the section modulus to be used.

Table 6-7 gives the properties of standard dressed sizes of structural lumber. Note the difference between the nominal and dressed dimensions. The properties A, I, and S are, of course, based on actual dressed sizes.

TABLE 6-1. Properties of Wide Flange (WF) Shapes

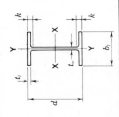

Designation	Area A in.²	Depth d in.	Web Thickness t_w in.	Flange Width t_f in.	Flange Thickness b_f in.	k in.	Axis X-X I in.⁴	Axis X-X S in.³	Axis X-X r in.	Axis Y-Y I in.⁴	Axis Y-Y S in.³	Axis Y-Y r in.
W 36 × 300	88.3	36.74	0.945	16.655	1.680	2.81	20,300	1110	15.2	1300	156	3.83
× 230	67.6	35.90	0.760	16.470	1.260	2.38	15,000	837	14.9	940	114	3.73
× 170	50.0	36.17	0.680	12.030	1.100	2.00	10,500	580	14.5	320	53.2	2.53
× 135	39.7	35.55	0.600	11.950	0.790	1.69	7,800	439	14.0	225	37.7	2.38
W 33 × 241	70.9	34.18	0.830	15.860	1.400	2.19	14,200	829	14.1	932	118	3.63
× 152	44.7	33.49	0.635	11.565	1.055	1.88	8,160	487	13.5	273	47.2	2.47
× 118	34.7	32.86	0.550	11.480	0.740	1.56	5,900	359	13.0	187	32.6	2.32
W 30 × 211	62.0	30.94	0.775	15.105	1.315	2.13	10,300	663	12.9	757	100	3.49
× 124	36.5	30.17	0.585	10.515	0.930	1.69	5,360	355	12.1	181	34.4	2.23
× 99	29.1	29.65	0.520	10.450	0.670	1.44	3,990	269	11.7	128	24.5	2.10
W 27 × 178	52.3	27.81	0.725	14.085	1.190	1.88	6,990	502	11.6	555	78.8	3.26
× 102	30.0	27.09	0.515	10.015	0.830	1.56	3,620	267	11.0	139	27.8	2.15
× 84	24.8	26.71	0.460	9.960	0.640	1.38	2,850	213	10.7	106	21.2	2.07

TABLE 6-1. (Continued)

Designation	Area A in.²	Depth d in.	Web Thickness t_w in.	Flange Width t_f in.	Flange Thickness b_f in.	k in.	Axis X-X I in.⁴	Axis X-X S in.³	Axis X-X r in.	Axis Y-Y I in.⁴	Axis Y-Y S in.³	Axis Y-Y r in.
W 24 × 162	47.7	25.00	0.705	12.955	1.220	2.00	5,170	414	10.4	443	68.4	3.05
× 104	30.6	24.06	0.500	12.750	0.750	1.50	3,100	258	10.1	259	40.7	2.91
× 84	24.7	24.10	0.470	9.020	0.770	1.56	2,370	196	9.79	94.4	20.9	1.95
× 55	16.2	23.57	0.395	7.005	0.505	1.94	1,350	114	9.11	29.1	8.30	1.34
W 21 × 147	43.2	22.06	0.720	12.510	1.150	1.88	3,630	329	9.17	376	60.1	2.95
× 101	29.8	21.36	0.500	12.290	0.800	1.56	2,420	227	9.02	248	40.3	2.89
× 57	16.7	21.06	0.405	6.555	0.650	1.38	1,170	111	8.36	30.6	9.35	1.35
× 44	13.0	20.66	0.350	6.500	0.450	1.19	834	81.6	8.06	20.7	6.36	1.26
W 18 × 119	35.1	18.97	0.655	11.265	1.060	1.75	2,190	231	7.90	253	44.9	2.69
× 76	22.3	18.21	0.425	11.035	0.680	1.38	1,330	146	7.73	152	27.6	2.61
× 55	16.2	18.11	0.390	7.530	0.630	1.31	890	98.3	7.41	44.9	11.9	1.67
× 46	13.5	18.06	0.360	6.060	0.605	1.25	712	78.8	7.25	22.5	7.43	1.29
× 35	10.3	17.70	0.300	6.000	0.425	1.13	510	57.6	7.04	15.3	5.12	1.22
W 16 × 100	29.4	16.97	0.585	10.425	0.985	1.69	1,490	175	7.10	186	35.7	2.51
× 57	16.8	16.43	0.430	7.120	0.715	1.38	758	92.2	6.72	43.1	12.1	1.60
× 45	13.3	16.13	0.345	7.035	0.565	1.25	586	72.7	6.65	32.8	9.34	1.57
× 36	10.6	15.86	0.295	6.985	0.430	1.13	448	56.5	6.51	24.5	7.00	1.52
× 26	7.68	15.69	0.250	5.500	0.345	1.06	301	38.4	6.26	9.59	3.49	1.12
W 14 × 730	215.0	22.42	3.070	17.890	4.910	5.56	14,300	1280	8.17	4720	527	4.69

× 605	178.0	20.92	2.595	17.415	4.160	4.81	10,800	1040	7.80	3680	423	4.55
× 370	109.0	17.92	1.655	16.475	2.660	3.31	5,440	607	7.07	1990	241	4.27
× 211	62.0	15.72	0.980	15.800	1.560	2.25	2,660	338	6.55	1030	130	4.07
× 120	35.3	14.48	0.590	14.670	0.940	1.63	1,380	190	6.24	495	67.5	3.74
× 90	26.5	14.02	0.440	14.520	0.710	1.38	999	143	6.14	362	49.9	3.70
× 68	20.0	14.04	0.415	10.035	0.720	1.50	723	103	6.01	121	24.2	2.46
× 43	12.6	13.66	0.305	7.995	0.530	1.31	428	62.7	5.82	45.2	11.3	1.89
× 30	8.85	13.84	0.270	6.730	0.385	0.94	291	42.0	5.73	19.6	5.82	1.49
× 22	6.49	13.74	0.230	5.000	0.335	0.88	199	29.0	5.54	7.00	2.80	1.04
W 12 × 336	98.8	16.82	1.775	13.385	2.955	3.69	4,060	483	6.41	1190	177	3.47
× 210	61.8	14.71	1.180	12.790	1.900	2.63	2,140	292	5.89	664	104	3.28
× 106	31.2	12.89	0.610	12.220	0.990	1.69	933	145	5.47	301	49.3	3.11
× 58	17.0	12.19	0.360	10.010	0.640	1.38	475	78.0	5.28	107	21.4	2.51
× 40	11.8	11.94	0.295	8.005	0.515	1.25	310	51.9	5.13	44.1	11.0	1.93
× 22	6.48	12.31	0.260	4.030	0.425	0.88	156	25.4	4.91	4.66	2.31	0.847
× 14	4.16	11.91	0.200	3.970	0.225	0.69	88.6	14.9	4.62	2.36	1.19	0.753
W 10 × 112	32.9	11.36	0.755	10.415	1.250	1.88	716	126	4.66	236	45.3	2.68
× 60	17.6	10.22	0.420	10.080	0.680	1.31	341	66.7	4.39	116	23.0	2.57
× 33	9.71	9.73	0.290	7.960	0.435	1.06	170	35.0	4.19	36.6	9.20	1.94
× 26	7.61	10.33	0.260	5.770	0.440	0.88	144	27.9	4.35	14.1	4.89	1.36
× 19	5.62	10.24	0.250	4.020	0.395	0.81	96.3	18.8	4.14	4.29	2.14	0.874
× 12	3.54	9.87	0.190	3.960	0.210	0.63	53.8	10.9	3.90	2.18	1.10	0.785
W 8 × 67	19.7	9.00	0.570	8.280	0.935	1.44	272	60.4	3.72	88.6	21.4	2.12
× 40	11.7	8.25	0.360	8.070	0.560	1.06	146	35.5	3.53	49.1	12.2	2.04
× 31	9.13	8.00	0.285	7.995	0.435	0.94	110	27.5	3.47	37.1	9.27	2.02
× 24	7.08	7.93	0.245	6.495	0.400	0.88	82.8	20.9	3.42	18.3	5.63	1.61

TABLE 6-1. (Continued)

Designation	Area A in.²	Depth d in.	Web Thickness t_w in.	Flange Width t_f in.	Flange Thickness b_f in.	k in.	Axis X-X I in.⁴	Axis X-X S in.³	Axis X-X r in.	Axis Y-Y I in.⁴	Axis Y-Y S in.³	Axis Y-Y r in.
W 8 × 18	5.26	8.14	0.230	5.250	0.330	0.75	61.9	15.2	3.43	7.97	3.04	1.23
× 13	3.84	7.99	0.230	4.000	0.255	0.69	39.6	9.91	3.21	2.73	1.37	0.843
× 10	2.96	7.89	0.170	3.940	0.205	0.63	30.8	7.81	3.22	2.09	1.06	0.841
W 6 × 25	7.34	6.38	0.320	6.080	0.455	0.81	53.4	16.7	2.70	17.1	5.61	1.52
× 15	4.43	5.99	0.230	5.990	0.260	0.63	29.1	9.72	2.56	9.32	3.11	1.46
× 12	3.55	6.03	0.230	4.000	0.280	0.63	22.1	7.31	2.49	2.99	1.50	0.918
× 9	2.68	5.90	0.170	3.940	0.215	0.56	16.4	5.56	2.47	2.19	1.11	0.905
W 5 × 19	5.54	5.15	0.270	5.030	0.430	0.81	26.2	10.2	2.17	9.13	3.63	1.28
× 16	4.68	5.01	0.240	5.000	0.360	0.75	21.3	8.51	2.13	7.51	3.00	1.27
W 4 × 13	3.83	4.16	0.280	4.060	0.345	0.69	11.3	5.46	1.72	3.86	1.90	1.00

Source: Adapted from data in the *Manual of Steel Construction*, 8th ed., with permission of the publishers, American Institute of Steel Construction.

TABLE 6-2. Properties of American Standard (S) Shapes (I-Beams)

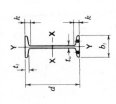

Designation	Area A in.²	Depth d in.	Web Thickness t_w in.	Flange Width t_f in.	Flange Thickness b_f in.	k in.	Axis X-X I in.⁴	Axis X-X S in.³	Axis X-X r in.	Axis Y-Y I in.⁴	Axis Y-Y S in.³	Axis Y-Y r in.
S 24 × 121	35.6	24.50	0.800	8.050	1.090	2.00	3160	258	9.43	83.3	20.7	1.53
× 106	31.2	24.50	0.620	7.870	1.090	2.00	2940	240	9.71	77.1	19.6	1.57
× 100	29.3	24.00	0.745	7.245	0.870	1.75	2390	199	9.02	47.7	13.2	1.27
× 90	26.5	24.00	0.625	7.125	0.870	1.75	2250	187	9.21	44.9	12.6	1.30
× 80	23.5	24.00	0.500	7.000	0.870	1.75	2100	175	9.47	42.2	12.1	1.34
S 20 × 96	28.2	20.30	0.800	7.200	0.920	1.75	1670	165	7.71	50.2	13.9	1.33
× 86	25.3	20.30	0.660	7.060	0.920	1.75	1580	155	7.89	46.8	13.3	1.36
× 75	22.0	20.00	0.635	6.385	0.795	1.63	1280	128	7.62	29.8	9.32	1.16
× 66	19.4	20.00	0.505	6.255	0.795	1.63	1190	119	7.83	27.7	8.85	1.19
S 18 × 70	20.6	18.00	0.711	6.251	0.691	1.50	926	103	6.71	24.1	7.72	1.08
× 54.7	16.1	18.00	0.461	6.001	0.691	1.50	804	89.4	7.07	20.8	6.94	1.14
S 15 × 50	14.7	15.00	0.550	5.640	0.622	1.38	486	64.8	5.75	15.7	5.57	1.03
× 42.9	12.6	15.00	0.411	5.501	0.622	1.38	447	59.6	5.95	14.4	5.23	1.07

Elastic Properties

TABLE 6-2. (Continued)

Designation	Area A in.^{2}	Depth d in.	Web Thickness t_w in.	Flange Width t_f in.	Flange Thickness b_f in.	k in.	Axis X-X I in.^{4}	S in.^{3}	r in.	Axis Y-Y I in.^{4}	S in.^{3}	r in.
S 12 × 50	14.7	12.00	0.687	5.477	0.659	1.44	305	50.8	4.55	15.7	5.74	1.03
× 40.8	12.0	12.00	0.462	5.252	0.659	1.44	272	45.4	4.77	13.6	5.16	1.06
× 35	10.3	12.00	0.428	5.078	0.544	1.19	229	38.2	4.72	9.87	3.89	0.980
× 31.8	9.35	12.00	0.350	5.000	0.544	1.19	218	36.4	4.83	9.36	3.74	1.00
S 10 × 35	10.3	10.00	0.594	4.944	0.491	1.13	147	29.4	3.78	8.36	3.38	0.901
× 25.4	7.46	10.00	0.311	4.661	0.491	1.13	124	24.7	4.07	6.79	2.91	0.954
S 8 × 23	6.77	8.00	0.441	4.171	0.426	1.00	64.9	16.2	3.10	4.31	2.07	0.798
× 18.4	5.41	8.00	0.271	4.001	0.426	1.00	57.6	14.4	3.26	3.73	1.86	0.831
S 7 × 20	5.88	7.00	0.450	3.860	0.392	0.94	42.4	12.1	2.69	3.17	1.64	0.734
× 15.3	4.50	7.00	0.252	3.662	0.392	0.94	36.7	10.5	2.86	2.64	1.44	0.766
S 6 × 17.25	5.07	6.00	0.465	3.565	0.359	0.88	26.3	8.77	2.28	2.31	1.30	0.675
× 12.5	3.67	6.00	0.232	3.332	0.359	0.88	22.1	7.37	2.45	1.82	1.09	0.705
S 5 × 14.75	4.34	5.00	0.494	3.284	0.326	0.81	15.2	6.09	1.87	1.67	1.01	0.620
× 10	2.94	5.00	0.214	3.004	0.326	0.81	12.3	4.92	2.05	1.22	0.809	0.643
S 4 × 9.5	2.79	4.00	0.326	2.796	0.293	0.75	6.79	3.39	1.56	0.903	0.646	0.569
× 7.7	2.26	4.00	0.193	2.663	0.293	0.75	6.08	3.04	1.64	0.764	0.574	0.581
S 3 × 7.5	2.21	3.00	0.349	2.509	0.260	0.69	2.93	1.95	1.15	0.586	0.468	0.516
× 5.7	1.67	3.00	0.170	2.330	0.260	0.69	2.52	1.68	1.23	0.455	0.390	0.522

Source: Adapted from data in the *Manual of Steel Construction*, 8th ed., with permission of the publishers, American Institute of Steel Construction.

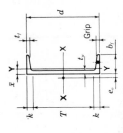

TABLE 6-3. Properties of American Standard Channel (C) Shapes

Designation	Area A in.²	Depth d in.	Web Thickness t_w in.	Flange Width t_f in.	Flange Thickness b_f in.	k in.	Axis X-X I in.⁴	S in.³	r in.	Axis Y-Y I in.⁴	S in.³	r in.	$\bar{x}$ in.
C 15 × 50	14.7	15.00	0.716	3.716	0.650	1.44	404	53.8	5.24	11.0	3.78	0.867	0.798
× 40	11.8	15.00	0.520	3.520	0.650	1.44	349	46.5	5.44	9.23	3.37	0.886	0.777
× 33.9	9.96	15.00	0.400	3.400	0.650	1.44	315	42.0	5.62	8.13	3.11	0.904	0.787
C 12 × 30	8.82	12.00	0.510	3.170	0.501	1.13	162	27.0	4.29	5.14	2.06	0.763	0.674
× 25	7.35	12.00	0.387	3.047	0.501	1.13	144	24.1	4.43	4.47	1.88	0.780	0.674
× 20.7	6.09	12.00	0.282	2.942	0.501	1.13	129	21.5	4.61	3.88	1.73	0.799	0.698
C 10 × 30	8.82	10.00	0.673	3.033	0.436	1.00	103	20.7	3.42	3.94	1.65	0.669	0.649
× 25	7.35	10.00	0.526	2.886	0.436	1.00	91.2	18.2	3.52	3.36	1.48	0.676	0.617
× 20	5.88	10.00	0.379	2.739	0.436	1.00	78.9	15.8	3.66	2.81	1.32	0.692	0.606
× 15.3	4.49	10.00	0.240	2.600	0.436	1.00	67.4	13.5	3.87	2.28	1.16	0.713	0.634
C 9 × 20	5.88	9.00	0.448	2.648	0.413	0.94	60.9	13.5	3.22	2.42	1.17	0.642	0.583
× 15	4.41	9.00	0.285	2.485	0.413	0.94	51.0	11.3	3.40	1.93	1.01	0.661	0.586
× 13.4	3.94	9.00	0.233	2.433	0.413	0.94	47.9	10.6	3.48	1.76	0.962	0.669	0.601

Elastic Properties

TABLE 6-3. (Continued)

Designation	Area A in.²	Depth d in.	Web Thickness t_w in.	Flange Width t_f in.	Flange Thickness b_f in.	k in.	Axis X-X I in.⁴	Axis X-X S in.³	Axis X-X r in.	Axis Y-Y I in.⁴	Axis Y-Y S in.³	Axis Y-Y r in.	$\bar{x}$ in.
C 8 × 18.75	5.51	8.00	0.487	2.527	0.390	0.94	44.0	11.0	2.82	1.98	1.01	0.599	0.565
× 13.75	4.04	8.00	0.303	2.343	0.390	0.94	36.1	9.03	2.99	1.53	0.854	0.615	0.553
× 11.5	3.38	8.00	0.220	2.260	0.390	0.94	32.6	8.14	3.11	1.32	0.781	0.625	0.571
C 7 × 14.75	4.33	7.00	0.419	2.299	0.366	0.88	27.2	7.78	2.51	1.38	0.779	0.564	0.532
× 12.25	3.60	7.00	0.314	2.194	0.366	0.88	24.2	6.93	2.60	1.17	0.703	0.571	0.525
× 9.8	2.87	7.00	0.210	2.090	0.366	0.88	21.3	6.08	2.72	0.968	0.625	0.581	0.540
C 6 × 13	3.83	6.00	0.437	2.157	0.343	0.81	17.4	5.80	2.13	1.05	0.642	0.525	0.514
× 10.5	3.09	6.00	0.314	2.034	0.343	0.81	15.2	5.06	2.22	0.866	0.564	0.529	0.499
× 8.2	2.40	6.00	0.200	1.920	0.343	0.81	13.1	4.38	2.34	0.693	0.492	0.537	0.511
C 5 × 9	2.64	5.00	0.325	1.885	0.320	0.75	8.90	3.56	1.83	0.632	0.450	0.489	0.478
× 6.7	1.97	5.00	0.190	1.750	0.320	0.75	7.49	3.00	1.95	0.479	0.378	0.493	0.484
C 4 × 7.25	2.13	4.00	0.321	1.721	0.296	0.69	4.59	2.29	1.47	0.433	0.343	0.450	0.459
× 5.4	1.59	4.00	0.184	1.584	0.296	0.69	3.85	1.93	1.56	0.319	0.283	0.449	0.457
C 3 × 6	1.76	3.00	0.356	1.596	0.273	0.69	2.07	1.38	1.08	0.305	0.268	0.416	0.455
× 5	1.47	3.00	0.258	1.498	0.273	0.69	1.85	1.24	1.12	0.247	0.233	0.410	0.438
× 4.1	1.21	3.00	0.170	1.410	0.273	0.69	1.66	1.10	1.17	0.197	0.202	0.404	0.436

Source: Adapted from data in the *Manual of Steel Construction*, 8th ed., with permission of the publishers, American Institute of Steel Construction.

TABLE 6-4. Properties of Angles

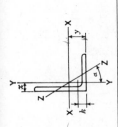

Size and Thickness in.	k in.	Weight lb	Area in.²	Axis X-X I in.⁴	S in.³	r in.	y in.	Axis Y-Y I in.⁴	S in.³	r in.	x in.	Axis Z-Z r in.	Tan α
8 × 8 × 1 1/8	1.75	56.9	16.7	98.0	17.5	2.42	2.41	98.0	17.5	2.42	2.41	1.56	1.000
1	1.625	51.0	15.0	89.0	15.8	2.44	2.37	89.0	15.8	2.44	2.37	1.56	1.000
7/8	1.50	45.0	13.2	79.6	14.0	2.45	2.32	79.6	14.0	2.45	2.32	1.57	1.000
3/4	1.375	38.9	11.4	69.7	12.2	2.47	2.28	69.7	12.2	2.47	2.28	1.58	1.000
5/8	1.25	32.7	9.61	59.4	10.3	2.49	2.23	59.4	10.3	2.49	2.23	1.58	1.000
1/2	1.125	26.4	7.75	48.6	8.36	2.50	2.19	48.6	8.36	2.50	2.19	1.59	1.000
8 × 6 × 1	1.50	44.2	13.0	80.8	15.1	2.49	2.65	38.8	8.92	1.73	1.65	1.28	0.543
3/4	1.25	33.8	9.94	63.4	11.7	2.53	2.56	30.7	6.92	1.76	1.56	1.29	0.551
1/2	1.0	23.0	6.75	44.3	8.02	2.56	2.47	21.7	4.79	1.79	1.47	1.30	0.558
8 × 4 × 1	1.50	37.4	11.0	69.6	14.1	2.52	3.05	11.6	3.94	1.03	1.05	0.846	0.247
3/4	1.25	28.7	8.44	54.9	10.9	2.55	2.95	9.36	3.07	1.05	0.953	0.852	0.258
1/2	1.0	19.6	5.75	38.5	7.49	2.59	2.86	6.74	2.15	1.08	0.859	0.865	0.267
7 × 4 × 3/4	1.25	26.2	7.69	37.8	8.42	2.22	2.51	9.05	3.03	1.09	1.01	0.860	0.324
1/2	1.0	17.9	5.25	26.7	5.81	2.25	2.42	6.53	2.12	1.11	0.917	0.872	0.335
3/8	0.875	13.6	3.98	20.6	4.44	2.27	2.37	5.10	1.63	1.13	0.870	0.880	0.340
6 × 6 × 1	1.50	37.4	11.0	35.5	8.57	1.80	1.86	35.5	8.57	1.80	1.86	1.17	1.000
7/8	1.375	33.1	9.73	31.9	7.63	1.81	1.82	31.9	7.63	1.81	1.82	1.17	1.000

TABLE 6-4. (Continued)

Size and Thickness in.	k in.	Weight lb	Area in.²	Axis X-X				Axis Y-Y				Axis Z-Z	
				I in.⁴	S in.³	r in.	y in.	I in.⁴	S in.³	r in.	x in.	r in.	Tan α
6 × 6 × 3/4	1.25	28.7	8.44	28.2	6.66	1.83	1.78	28.2	6.66	1.83	1.78	1.17	1.000
5/8	1.125	24.2	7.11	24.2	5.66	1.84	1.73	24.2	5.66	1.84	1.73	1.18	1.000
1/2	1.0	19.6	5.75	19.9	4.61	1.86	1.68	19.9	4.61	1.86	1.68	1.18	1.000
3/8	0.875	14.9	4.36	15.4	3.53	1.88	1.64	15.4	3.53	1.88	1.64	1.19	1.000
6 × 4 × 3/4	1.25	23.6	6.94	24.5	6.25	1.88	2.08	8.68	2.97	1.12	1.08	0.860	0.428
5/8	1.125	20.0	5.86	21.1	5.31	1.90	2.03	7.52	2.54	1.13	1.03	0.864	0.435
1/2	1.00	16.2	4.75	17.4	4.33	1.91	1.99	6.27	2.08	1.15	0.987	0.870	0.440
3/8	0.875	12.3	3.61	13.5	3.32	1.93	1.94	4.90	1.60	1.17	0.941	0.877	0.446
6 × 3 1/2 × 3/8	0.875	11.7	3.42	12.9	3.24	1.94	2.04	3.34	1.23	0.988	0.787	0.767	0.350
5/16	0.8125	9.8	2.87	10.9	2.73	1.95	2.01	2.85	1.04	0.996	0.763	0.772	0.352
5 × 5 × 7/8	1.375	27.2	7.98	17.8	5.17	1.49	1.57	17.8	5.17	1.49	1.57	0.973	1.000
3/4	1.25	23.6	6.94	15.7	4.53	1.51	1.52	15.7	4.53	1.51	1.52	0.975	1.000
1/2	1.0	16.2	4.75	11.3	3.16	1.54	1.43	11.3	3.16	1.54	1.43	0.983	1.000
5 × 5 × 3/8	0.875	12.3	3.61	8.74	2.42	1.56	1.39	8.74	2.42	1.56	1.39	0.990	1.000
5/16	0.8125	10.3	3.03	7.42	2.04	1.57	1.37	7.42	2.04	1.57	1.37	0.994	1.000
5 × 3 1/2 × 3/4	1.25	19.8	5.81	13.9	4.28	1.55	1.75	5.55	2.22	0.977	0.996	0.748	0.464
1/2	1.0	13.6	4.00	9.99	2.99	1.58	1.66	4.05	1.56	1.01	0.906	0.755	0.479
3/8	0.875	10.4	3.05	7.78	2.29	1.60	1.61	3.18	1.21	1.02	0.861	0.762	0.486
5/16	0.8125	8.7	2.56	6.60	1.94	1.61	1.59	2.72	1.02	1.03	0.838	0.766	0.489
5 × 3 × 1/2	1.0	12.8	3.75	9.45	2.91	1.59	1.75	2.58	1.15	0.829	0.750	0.648	0.357
3/8	0.875	9.8	2.86	7.37	2.24	1.61	1.70	2.04	0.888	0.845	0.704	0.654	0.364
5/16	0.8125	8.2	2.40	6.26	1.89	1.61	1.68	1.75	0.753	0.853	0.681	0.658	0.368
1/4	0.75	6.6	1.94	5.11	1.53	1.62	1.66	1.44	0.614	0.861	0.657	0.663	0.371

Size	k	Wt.	Area	I_x	S_x	r_x	y	I_y	S_y	r_y	x	r_z	tan α
4 × 4 × 3/4	1.125	18.5	5.44	7.67	2.81	1.19	1.27	7.67	2.81	1.19	1.27	0.778	1.000
5/8	1.0	15.7	4.61	6.66	2.40	1.20	1.23	6.66	2.40	1.20	1.23	0.779	1.000
1/2	0.875	12.8	3.75	5.56	1.97	1.22	1.18	5.56	1.97	1.22	1.18	0.782	1.000
3/8	0.75	9.8	2.86	4.36	1.52	1.23	1.14	4.36	1.52	1.23	1.14	0.788	1.000
5/16	0.6875	8.2	2.40	3.71	1.29	1.24	1.12	3.71	1.29	1.24	1.12	0.791	1.000
1/4	0.625	6.6	1.94	3.04	1.05	1.25	1.09	3.04	1.05	1.25	1.09	0.795	1.000
4 × 3 1/2 × 1/2	0.9375	11.9	3.50	5.32	1.94	1.23	1.25	3.79	1.52	1.04	1.00	0.722	0.750
3/8	0.8125	9.1	2.67	4.18	1.49	1.25	1.21	2.95	1.17	1.06	0.955	0.727	0.755
5/16	0.75	7.7	2.25	3.56	1.26	1.26	1.18	2.55	0.994	1.07	0.932	0.730	0.757
1/4	0.6875	6.2	1.81	2.91	1.03	1.27	1.16	2.09	0.808	1.07	0.909	0.734	0.759
4 × 3 × 1/2	0.9375	11.1	3.25	5.05	1.89	1.25	1.33	2.42	1.12	0.864	0.827	0.639	0.543
3/8	0.8125	8.5	2.48	3.96	1.46	1.26	1.28	1.92	0.866	0.879	0.782	0.644	0.551
5/16	0.75	7.2	2.09	3.38	1.23	1.27	1.26	1.65	0.734	0.887	0.759	0.647	0.554
1/4	0.6875	5.8	1.69	2.77	1.00	1.28	1.24	1.36	0.599	0.896	0.736	0.651	0.558
3 1/2 × 3 1/2 × 3/8	0.75	8.5	2.48	2.87	1.15	1.07	1.01	2.87	1.15	1.07	1.01	0.687	1.000
5/16	0.6875	7.2	2.09	2.45	0.976	1.08	0.990	2.45	0.976	1.08	0.990	0.690	1.000
1/4	0.625	5.8	1.69	2.01	0.794	1.09	0.968	2.01	0.794	1.09	0.968	0.694	1.000
3 1/2 × 3 × 3/8	0.8125	7.9	2.30	2.72	1.13	1.09	1.08	1.85	0.851	0.897	0.830	0.625	0.721
5/16	0.75	6.6	1.93	2.33	0.954	1.10	1.06	1.58	0.722	0.905	0.808	0.627	0.724
1/4	0.6875	5.4	1.56	1.91	0.776	1.11	1.04	1.30	0.589	0.914	0.785	0.631	0.727
3 1/2 × 2 1/2 × 3/8	0.8125	7.2	2.11	2.56	1.09	1.10	1.16	1.09	0.592	0.719	0.660	0.537	0.496
5/16	0.75	6.1	1.78	2.19	0.927	1.11	1.14	0.939	0.504	0.727	0.637	0.540	0.501
1/4	0.6875	4.9	1.44	1.80	0.755	1.12	1.11	0.777	0.412	0.735	0.614	0.544	0.506
3 × 3 × 1/2	0.8125	9.4	2.75	2.22	1.07	0.898	0.932	2.22	1.07	0.898	0.932	0.584	1.000
3/8	0.6875	7.2	2.11	1.76	0.833	0.913	0.888	1.76	0.833	0.913	0.888	0.587	1.000
5/16	0.625	6.1	1.78	1.51	0.707	0.922	0.865	1.51	0.707	0.922	0.865	0.589	1.000
1/4	0.5625	4.9	1.44	1.24	0.577	0.930	0.842	1.24	0.577	0.930	0.842	0.592	1.000
3/16	0.5000	3.71	1.09	0.962	0.441	0.939	0.820	0.962	0.441	0.939	0.820	0.596	1.000

TABLE 6-4. (Continued)

Size and Thickness in.	k in.	Weight lb	Area in.²	Axis X-X				Axis Y-Y				Axis Z-Z	
				I in.⁴	S in.³	r in.	y in.	I in.⁴	S in.³	r in.	x in.	r in.	Tan α
3 × 2 1/2 × 3/8	0.75	6.6	1.92	1.66	0.810	0.928	0.956	1.04	0.581	0.736	0.706	0.522	0.676
1/4	0.625	4.5	1.31	1.17	0.561	0.945	0.911	0.743	0.404	0.753	0.661	0.528	0.684
3/16	0.5625	3.39	0.996	0.907	0.430	0.954	0.888	0.577	0.310	0.761	0.638	0.533	0.688
3 × 2 × 3/8	0.6875	5.9	1.73	1.53	0.781	0.940	1.04	0.543	0.371	0.559	0.539	0.430	0.428
5/16	0.625	5.0	1.46	1.32	0.664	0.948	1.02	0.470	0.317	0.567	0.516	0.432	0.435
1/4	0.5625	4.1	1.19	1.09	0.542	0.957	0.993	0.392	0.260	0.574	0.493	0.435	0.440
3/16	0.5000	3.07	0.902	0.842	0.415	0.966	0.970	0.307	0.200	0.583	0.470	0.439	0.446
2 1/2 × 2 1/2 × 3/8	0.6875	5.9	1.73	0.984	0.566	0.753	0.762	0.984	0.566	0.753	0.762	0.487	1.000
5/16	0.625	5.0	1.46	0.849	0.482	0.761	0.740	0.849	0.482	0.761	0.740	0.489	1.000
1/4	0.5625	4.1	1.19	0.703	0.394	0.769	0.717	0.703	0.394	0.769	0.717	0.491	1.000
3/16	0.5000	3.07	0.902	0.547	0.303	0.778	0.694	0.547	0.303	0.778	0.694	0.495	1.000
2 1/2 × 2 × 3/8	0.6875	5.3	1.55	0.912	0.547	0.768	0.831	0.514	0.363	0.577	0.581	0.420	0.614
5/16	0.625	4.5	1.31	0.788	0.466	0.776	0.809	0.446	0.310	0.584	0.559	0.422	0.620
1/4	0.5625	3.62	1.06	0.654	0.381	0.784	0.787	0.372	0.254	0.592	0.537	0.424	0.626
3/16	0.5000	2.75	0.809	0.509	0.293	0.793	0.764	0.291	0.196	0.600	0.514	0.427	0.631
2 × 2 × 3/8	0.6875	4.7	1.36	0.479	0.351	0.594	0.636	0.479	0.351	0.594	0.636	0.389	1.000
5/16	0.625	3.92	1.15	0.416	0.300	0.601	0.614	0.416	0.300	0.601	0.614	0.390	1.000
1/4	0.5625	3.19	0.938	0.348	0.247	0.609	0.592	0.348	0.247	0.609	0.592	0.391	1.000
3/16	0.5000	2.44	0.715	0.272	0.190	0.617	0.569	0.272	0.190	0.617	0.569	0.394	1.000

Source: Abstracted from the *Manual of Steel Construction*, 8th ed., with permission of the publishers, American Institute of Steel Construction.

TABLE 6-5. Properties of Round Steel Pipe

Dimensions				Weight per Foot Lbs. Plain Ends	Properties			
Nominal Diameter In.	Outside Diameter In.	Inside Diameter In.	Wall Thickness In.		A In.^{2}	I In.^{4}	S In.^{3}	r In.
Standard Weight								
½	.840	.622	.109	.85	.250	.017	.041	.261
¾	1.050	.824	.113	1.13	.333	.037	.071	.334
1	1.315	1.049	.133	1.68	.494	.087	.133	.421
1¼	1.660	1.380	.140	2.27	.669	.195	.235	.540
1½	1.900	1.610	.145	2.72	.799	.310	.326	.623
2	2.375	2.067	.154	3.65	1.07	.666	.561	.787
2½	2.875	2.469	.203	5.79	1.70	1.53	1.06	.947
3	3.500	3.068	.216	7.58	2.23	3.02	1.72	1.16
3½	4.000	3.548	.226	9.11	2.68	4.79	2.39	1.34
4	4.500	4.026	.237	10.79	3.17	7.23	3.21	1.51
5	5.563	5.047	.258	14.62	4.30	15.2	5.45	1.88
6	6.625	6.065	.280	18.97	5.58	28.1	8.50	2.25
8	8.625	7.981	.322	28.55	8.40	72.5	16.8	2.94
10	10.750	10.020	.365	40.48	11.9	161	29.9	3.67
12	12.750	12.000	.375	49.56	14.6	279	43.8	4.38
Extra Strong								
½	.840	.546	.147	1.09	.320	.020	.048	.250
¾	1.050	.742	.154	1.47	.433	.045	.085	.321
1	1.315	.957	.179	2.17	.639	.106	.161	.407
1¼	1.660	1.278	.191	3.00	.881	.242	.291	.524
1½	1.900	1.500	.200	3.63	1.07	.391	.412	.605
2	2.375	1.939	.218	5.02	1.48	.868	.731	.766
2½	2.875	2.323	.276	7.66	2.25	1.92	1.34	.924
3	3.500	2.900	.300	10.25	3.02	3.89	2.23	1.14
3½	4.000	3.364	.318	12.50	3.68	6.28	3.14	1.31
4	4.500	3.826	.337	14.98	4.41	9.61	4.27	1.48
5	5.563	4.813	.375	20.78	6.11	20.7	7.43	1.84
6	6.625	5.761	.432	28.57	8.40	40.5	12.2	2.19
8	8.625	7.625	.500	43.39	12.8	106	24.5	2.88
10	10.750	9.750	.500	54.74	16.1	212	39.4	3.63
12	12.750	11.750	.500	65.42	19.2	362	56.7	4.33
Double-Extra Strong								
2	2.375	1.503	.436	9.03	2.66	1.31	1.10	.703
2½	2.875	1.771	.552	13.69	4.03	2.87	2.00	.844
3	3.500	2.300	.600	18.58	5.47	5.99	3.42	1.05
4	4.500	3.152	.674	27.54	8.10	15.3	6.79	1.37
5	5.563	4.063	.750	38.55	11.3	33.6	12.1	1.72
6	6.625	4.897	.864	53.16	15.6	66.3	20.0	2.06
8	8.625	6.875	.875	72.42	21.3	162	37.6	2.76

The listed sections are available in conformance with ASTM Specification A53 Grade B or A501. Other sections are made to these specifications. Consult with pipe manufacturers or distributors for availability.

TABLE 6-6. Properties of Structural Steel Tubing

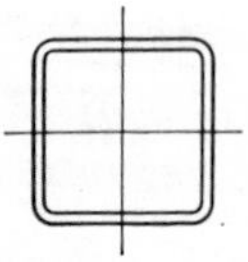

Square

DIMENSIONS				PROPERTIES**			
Nominal* Size	Wall Thickness		Weight per Foot	Area	I	S	r
In.	In.		Lb.	In.²	In.⁴	In.³	In.
16 x 16	.5000	½	103.30	30.4	1200	150	6.29
	.3750	⅜	78.52	23.1	931	116	6.35
	.3125	⁵/₁₆	65.87	19.4	789	98.6	6.38
14 x 14	.5000	½	89.68	26.4	791	113	5.48
	.3750	⅜	68.31	20.1	615	87.9	5.54
	.3125	⁵/₁₆	57.36	16.9	522	74.6	5.57
12 x 12	.5000	½	76.07	22.4	485	80.9	4.66
	.3750	⅜	58.10	17.1	380	63.4	4.72
	.3125	⁵/₁₆	48.86	14.4	324	54.0	4.75
	.2500	¼	39.43	11.6	265	44.1	4.78
10 x 10	.6250	⅝	76.33	22.4	321	64.2	3.78
	.5000	½	62.46	18.4	271	54.2	3.84
	.3750	⅜	47.90	14.1	214	42.9	3.90
	.3125	⁵/₁₆	40.35	11.9	183	36.7	3.93
	.2500	¼	32.63	9.59	151	30.1	3.96
8 x 8	.6250	⅝	59.32	17.4	153	38.3	2.96
	.5000	½	48.85	14.4	131	32.9	3.03
	.3750	⅜	37.69	11.1	106	26.4	3.09
	.3125	⁵/₁₆	31.84	9.36	90.9	22.7	3.12
	.2500	¼	25.82	7.59	75.1	18.8	3.15
	.1875	³/₁₆	19.63	5.77	58.2	14.6	3.18
7 x 7	.5000	½	42.05	12.4	84.6	24.2	2.62
	.3750	⅜	32.58	9.58	68.7	19.6	2.68
	.3125	⁵/₁₆	27.59	8.11	59.5	17.0	2.71
	.2500	¼	22.42	6.59	49.4	14.1	2.74
	.1875	³/₁₆	17.08	5.02	38.5	11.0	2.77
6 x 6	.5000	½	35.24	10.4	50.5	16.8	2.21
	.3750	⅜	27.48	8.08	41.6	13.9	2.27
	.3125	⁵/₁₆	23.34	6.86	36.3	12.1	2.30
	.2500	¼	19.02	5.59	30.3	10.1	2.33
	.1875	³/₁₆	14.53	4.27	23.8	7.93	2.36
5 x 5	.5000	½	28.43	8.36	27.0	10.8	1.80
	.3750	⅜	22.37	6.58	22.8	9.11	1.86
	.3125	⁵/₁₆	19.08	5.61	20.1	8.02	1.89
	.2500	¼	15.62	4.59	16.9	6.78	1.92
	.1875	³/₁₆	11.97	3.52	13.4	5.36	1.95

 * Outside dimensions across flat sides.
 ** Properties are based upon a nominal outside corner radius equal to two times the wall thickness.

TABLE 6-7. Properties of Structural Lumber

Dimensions (in.)		Area	Section Modulus	Moment of Inertia	
Nominal b h	Actual b h	A in.^{2}	S in.^{3}	I in.^{4}	Weight[a] lb/ft
2 × 3	1.5 × 2.5	3.75	1.563	1.953	0.9
2 × 4	1.5 × 3.5	5.25	3.063	5.359	1.3
2 × 6	1.5 × 5.5	8.25	7.563	20.797	2.0
2 × 8	1.5 × 7.25	10.875	13.141	47.635	2.6
2 × 10	1.5 × 9.25	13.875	21.391	98.932	3.4
2 × 12	1.5 × 11.25	16.875	31.641	177.979	4.1
2 × 14	1.5 × 13.25	19.875	43.891	290.775	4.8
3 × 2	2.5 × 1.5	3.75	0.938	0.703	0.9
3 × 4	2.5 × 3.5	8.75	5.104	8.932	2.1
3 × 6	2.5 × 5.5	13.75	12.604	34.661	3.3
3 × 8	2.5 × 7.25	18.125	21.901	79.391	4.4
3 × 10	2.5 × 9.25	23.125	35.651	164.886	5.6
3 × 12	2.5 × 11.25	28.125	52.734	296.631	6.8
3 × 14	2.5 × 13.25	33.125	73.151	484.625	8.1
3 × 16	2.5 × 15.25	38.125	96.901	738.870	9.3
4 × 2	3.5 × 1.5	5.25	1.313	0.984	1.3
4 × 3	3.5 × 2.5	8.75	3.646	4.557	2.1
4 × 4	3.5 × 3.5	12.25	7.146	12.505	3.0
4 × 6	3.5 × 5.5	19.25	17.646	48.526	4.7
4 × 8	3.5 × 7.25	23.375	30.661	111.148	6.2
4 × 10	3.5 × 9.25	32.375	49.911	230.840	7.9
4 × 12	3.5 × 11.25	39.375	73.828	415.283	9.6
4 × 14	3.5 × 13.25	46.375	102.411	678.475	11.3
4 × 16	3.5 × 15.25	53.375	135.661	1034.418	13.0
6 × 2	5.5 × 1.5	8.25	2.063	1.547	2.0
6 × 3	5.5 × 2.5	13.75	5.729	7.161	3.3
6 × 4	5.5 × 3.5	19.25	11.229	19.651	4.7
6 × 6	5.5 × 5.5	30.25	27.729	76.255	7.4
6 × 8	5.5 × 7.5	41.25	51.563	193.359	10.0
6 × 10	5.5 × 9.5	52.25	82.729	392.963	12.7
6 × 12	5.5 × 11.5	63.25	121.229	697.068	15.4
6 × 14	5.5 × 13.5	74.25	167.063	1127.672	18.0
6 × 16	5.5 × 15.5	85.25	220.229	1706.776	20.7
8 × 2	7.25 × 1.5	10.875	2.719	2.039	2.6
8 × 3	7.25 × 2.5	18.125	7.552	9.440	4.4
8 × 4	7.25 × 3.5	25.375	14.802	25.904	6.2

TABLE 6-7. (*Continued*)

Dimensions (in.)		Area A in.^{2}	Section Modulus S in.^{3}	Moment of Inertia I in.^{4}	Weight[a] lb/ft
Nominal b h	Actual b h				
8 × 6	7.5 × 5.5	41.25	37.813	103.984	10.0
8 × 8	7.5 × 7.5	56.25	70.313	263.672	13.7
8 × 10	7.5 × 9.5	71.25	112.813	535.859	17.3
8 × 12	7.5 × 11.5	86.25	165.313	950.547	21.0
8 × 14	7.5 × 13.5	101.25	227.813	1537.734	24.6
8 × 16	7.5 × 15.5	116.25	300.313	2327.422	28.3
8 × 18	7.5 × 17.5	131.25	382.813	3349.609	31.9
8 × 20	7.5 × 19.5	146.25	475.313	4634.297	35.5
10 × 10	9.5 × 9.5	90.25	142.896	678.755	21.9
10 × 12	9.5 × 11.5	109.25	209.396	1204.026	26.6
10 × 14	9.5 × 13.5	128.25	288.563	1947.797	31.2
10 × 16	9.5 × 15.5	147.25	380.396	2948.068	35.8
10 × 18	9.5 × 17.5	166.25	484.896	4242.836	40.4
10 × 20	9.5 × 19.5	185.25	602.063	5870.109	45.0
12 × 12	11.5 × 11.5	132.25	253.479	1457.505	32.1
12 × 14	11.5 × 13.5	155.25	349.313	2357.859	37.7
12 × 16	11.5 × 15.5	178.25	460.479	3568.713	43.3
12 × 18	11.5 × 17.5	201.25	586.979	5136.066	48.9
12 × 20	11.5 × 19.5	224.25	728.813	7105.922	54.5
12 × 22	11.5 × 21.5	247.25	885.979	9524.273	60.1
12 × 24	11.5 × 23.5	270.25	1058.479	12437.129	65.7
14 × 14	13.5 × 13.5	182.25	410.063	2767.922	44.3
16 × 16	15.5 × 15.5	240.25	620.646	4810.004	58.4

Source: Compiled from data in the *National Design Specification for Wood Construction*, 1982 ed., with permission of the publishers, National Forest Products Association.
[a] Based on an assumed average weight of 35 lb/ft^3.

7

Behavior of Beams: Reactions and Shears
IIIIIIIIIIIIIIIIIIIIIIIIIIIIIIIIIIIIIIIIIIIIIIII

7-1. Types of Beams

A beam is a structural member resting on supports, usually at its ends, that supports transverse loads. The loads on a beam tend to *bend* rather than to shorten or lengthen the member. The great majority of beams used in building construction are placed in a horizontal position. Generally, the loads on a beam that result from the force of gravity are vertical. The forces a beam is required to resist are the downward loads and the upward supporting forces called *reactions*.

Several types of beam may be identified by the number, kind, and position of the supports. In Figure 7-1 a number of common beams are shown with the exaggerated shape that each beam assumes when loaded.

A *simple beam* rests on a support at each end, the beam ends being free to rotate. A large percentage of the beams in steel-frame buildings are designed as simple beams (Figure 7-1a). When there is no restraint against rotation at the supports, we say that the beam is *simply supported*.

A *cantilever beam* projects beyond its support. A beam embedded in a wall and extending beyond the face of the wall is a typical ex-

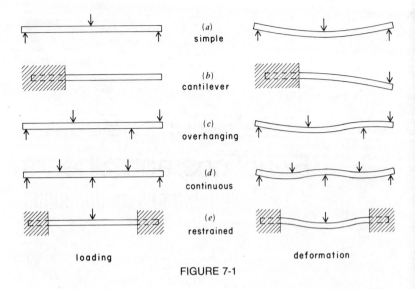

loading   FIGURE 7-1   *deformation*

ample. This is illustrated in Figure 7-1*b*. This beam is said to be *fixed* or *restrained* at the support.

An *overhanging beam* is a beam whose end or ends project beyond its supports. Figure 7-1*c* indicates an overhanging beam. The projecting end is a cantilever in which the stresses are similar to those in the beam shown in Figure 7-1*b*. It is shown later, however, that the stresses in the portion of the beam between the supports are different from the stresses in the simple beam shown in Figure 7-1*a*.

A *continuous beam* is supported on more than two supports (Figure 7-1*d*). Continuous beams are frequently used in welded steel framing and reinforced concrete construction.

A *restrained beam* has one or both ends restrained or *fixed* against rotation (Figure 7-1*e*). Fixed-end conditions occur when the end of a beam is rigidly connected to its supporting member.

When designing beams it is advantageous to make a diagram to show the loading conditions and to sketch directly below it the shape the beam tends to assume when it bends. The ability to visualize this deformation curve is important because, as we demonstrate later, the curve indicates changes in the character of the bending stresses along the beam's length.

7-2.   Loading

The loads supported by beams are classified as *concentrated* or *distributed*. A concentrated load is one that extends over so small a portion of the beam length that it may be assumed to act at a point, as indicated in the diagrams of Figure 7-1. A girder in a building takes concentrated loads at the points at which the floor beams frame into it (Figure 7-2). Also, the load exerted by a column that rests on a beam or girder is a concentrated load. Actually this load extends over a short length of the beam, represented by the width of the column base. For practical purposes, however, the load acts at the axis of the column. Similarly, a beam resting on a masonry wall produces a downward force that is resisted by the upward reaction. The reaction is distributed over the length of the beam (or its bearing plate) that rests on the wall, but the reaction may be considered to act at the midpoint of the beam's bearing area on the wall.

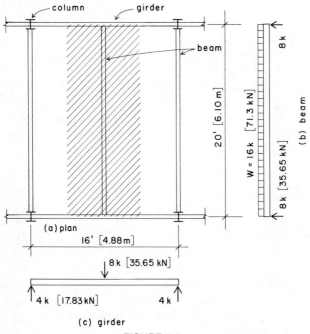

FIGURE 7-2

A distributed load is one that extends over a substantial portion or the entire length of a beam. Most distributed loads are *uniformly distributed*; that is, they have a uniform magnitude for each unit of length, such as pounds per linear foot, kips per linear foot, or kilonewtons per linear meter.

Figure 7-2*a* shows a portion of a floor framing plan. The diagonal crosshatching represents the area supported by one of the beams. This area is 8 ft [2.44 m] (the sum of one-half the distances to the next beam on each side), multiplied by the span length of 20 ft [6.10 m]. The beam is supported at each end by girders that span between the columns. Imagine that the floor load is 100 lb per sq ft [4.79 kN/m^2], including the weight of the construction. The total load on the beam is then $8 \times 20 \times 100 = 16,000$ lb, or 16 kips [$2.44 \times 6.10 \times 4.79 = 71.3$ kN]. It is common to designate this total load as W. Another way to determine the load is to say that it is $8 \times 100 = 800$ lb per ft [$2.44 \times 4.79 = 11.69$ kN/m], in which case it would be designated as w to indicate that it is a *unit load* rather than the total load on the beam. Most distributed loads are uniform and are constant over the entire length of the beam, as in this example. With different framing arrangements, however, the distributed loads may sometimes not be uniform or may extend over a limited portion of the beam length.

Because the beam in Figure 7-2 is symmetrically loaded, both end reactions are the same and are equal to one-half the total load on the beam. From this single beam each girder thus receives a concentrated load of one-half the beam load at the center of the girder span. Figures 7-2*b* and 7-2*c* are the conventional representations of the beam and girder load diagrams, respectively.

7-3. Vertical Shear

We may define *vertical shear* (designated V) as the tendency for one part of a member to move vertically with respect to an adjacent part. Referring to Figure 7-3, if the total uniformly distributed load on the beam is W, the loading is symmetrical and each reaction is $W/2$. If we imagine a vertical plane cutting a section through the beam flush with the face of the left supporting wall, the reaction of the wall on the beam to the left of the section is upward. Just to the right of the section the load on the beam acts downward, and the magnitude of the tendency

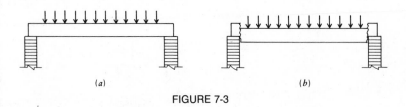

(a)   (b)

FIGURE 7-3

for the left and right portions to slide past each other (the vertical shear) is equal to the value of the reaction $W/2$ lb. Thus we say that at a section through the beam infinitely close to the support $V = W/2$ lb.

At any other vertical section taken along the beam span this same tendency exists for the portion to the left of the section to slip past the portion to the right. Therefore we may formulate the following definition:

The magnitude of the vertical shear at any section of a beam is equal to the algebraic sum of all the vertical forces on one side of the section.

To understand how readily the value of the vertical shear may be computed for any section of a beam let us call the upward forces (the reactions) positive and downward forces (the loads) negative. For the present let us consider only the forces to the *left* of the section. Then, in accordance with the foregoing definition, we can say that *the magnitude of the vertical shear at any section of a beam is equal to the sum of the reactions minus the sum of the loads of the left of the section.* To find the value of the shear at a particular section we simply note the forces to the left and then write shear = reactions − loads. Of course, we could consider the forces to the right of the section and find that the magnitude of the shear is the same, but in the following examples we consider the forces on the left. This procedure will avoid confusion with respect to algebraic sign, which is discussed later.

Let us try this rule on the beam shown in Figure 7-4. We find that the left reaction is 450 lb [2.0 kN] and that the right reaction is 1350 lb [6.0 kN]. When writing an equation for computing the value of the vertical shear it is convenient to identify the section at which the shear is taken by using a subscript: $V_{(x=4)}$. For this illustration this terminology indicates that the value of the vertical shear is taken at 4 ft [1.2 m] from the left end of the beam.

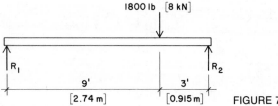

FIGURE 7-4

Now referring to Figure 7-4 and remembering the values computed for the reactions, let us write the value of the shear at 2 ft [0.6 m] from R_1. Repeat the foregoing rule and observe that the only force to the left of the section is the left reaction. Then

$$\text{shear} = \text{reaction} - \text{loads}$$

or

$$V_{(x=2)} = 450 - 0$$

Thus

$$V_{(x=2)} = 450 \text{ lb}$$

$$[V_{(x=0.6)} = 2 \text{ kN}]$$

Because there are no loads up to the load of 1800 lb [8 kN], the value of the shear in the beam is the same magnitude at any section in this portion of the beam. (From $x = 0$ to $x = 9$ ft.) [From $x = 0$ to $x = 2.7$ m.] Note, however, that we are ignoring the weight of the beam.

Next, consider a section 10 ft [3.0 m] from R_1. The reaction to the left of this section is 450 lb [2 kN], and the load to the left is 1800 lb [8 kN]. Then, because the shear equals the reactions minus the loads,

$$V_{(x=10)} = 450 - 1800 = -1350 \text{ lb}$$

$$[V_{(x=3.0)} = 2 - 8 = -6 \text{ kN}]$$

Because this beam has but one load, the value of the shear at any section between the load and the right reaction is -1350 lb [6 kN]. It should be noted here that for *simple* beams the value of the shear at the support is equal to the magnitude of the reaction. Hence the maximum shear will have the value of the greater reaction.

We compute the value of the shear in beams for two principal reasons. First, we must know the value of the maximum shear to be sure that there is ample material in the beam to prevent it from failing by shear. The second reason for computing the shear values is that *the greatest tendency for the beam to fail by bending is at the section of the beam at which the value of the shear is zero.* In the foregoing illustration we found both positive and negative values for the shear, and we also found that values on the left of the 1800 lb [8 kN] load were positive and those on the right were negative. The critical section for bending is under the concentrated load, where the shear changes from positive to negative. This is explained further when shear diagrams are constructed.

7-4. Shear Diagrams

In the preceding section we saw that the value of the shear may be computed readily at any section along the beam's length. When designing beams it will be found that diagrams that show the variation in shear along the span are extremely helpful.

To construct a shear diagram first make a drawing of the beam to scale, showing the loads and their positions. Next, compute the reactions as explained in Section 4-6. Below the beam draw a horizontal *base line* to represent zero shear. Then the values of the shear at various sections of the beam may be computed and plotted to a suitable scale vertically from the base line, positive values above and negative values below. The following examples illustrate the construction of shear diagrams for beams under various conditions of loading:

Example 1.   A simple beam 15 ft [4.5 m] long has two concentrated loads located as shown in Figure 7-5. Construct the shear diagram and note the value of the maximum shear and the section of the beam at which the shear passes through zero.

Solution: Taking R_2 as the center of moments,

$$15R_1 = (8000 \times 12) + (12,000 \times 5)$$

$$15R_1 = 156,000 \quad \text{and} \quad R_1 = 10,400 \text{ lb}$$

$$[4.5R_1 = (35.6 \times 3.6) + (53.4 \times 1.5)]$$

$$[4.5R_1 = 208.26 \quad \text{and} \quad R_1 = 46.28 \text{ kN}]$$

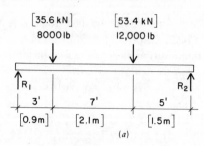

(a)

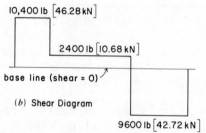

base line (shear = 0)

(b) Shear Diagram

9600 lb $\left[42.72\,\text{kN}\right]$ FIGURE 7-5

With R_1 as the center of moments,

$$15R_2 = (8000 \times 3) + (12,000 \times 10)$$

$$15R_2 = 144,000 \quad \text{and} \quad R_2 = 9600 \text{ lb}$$

$$[4.5R_2 = (35.6 \times 0.9) + (53.4 \times 3.0)]$$

$$[4.5R_2 = 192.24 \quad \text{and} \quad R_2 = 42.72 \text{ kN}]$$

Now let us compute the value of the shear at various sections:

$$V_{(x = 1)} = 10,400 - 0 = 10,400 \text{ lb}$$

$$V_{(x = 4)} = 10,400 - 8000 = 2400 \text{ lb}$$

$$V_{(x = 11)} = 10,400 - (8000 + 12,000) = -9600 \text{ lb}$$

$$\left[\begin{array}{c} V_{(x = 0.3)} = 46.28 - 0 = 46.28 \text{ kN} \\ V_{(x = 1.2)} = 46.28 - 35.6 = 10.68 \text{ kN} \\ V_{(x = 3.3)} = 46.28 - (35.6 + 53.4) = -42.72 \text{ kN} \end{array} \right]$$

Because there are only two concentrated loads, it is unnecessary to compute the shear for other sections; for instance, the value of the shear at any section between the two loads is 2400 lb [10.68 kN].

We may now construct the shear diagram shown in Figure 7-5 by plotting the three values of shear just computed, using the base line as the reference for zero shear. Positive values are plotted above the base line and negative values below the base line. Observing from the beam and its loads that the shear is a constant value between the points of location of the reactions and the loads, we can complete the shear diagram by drawing horizontal lines through the plotted points.

The value of the maximum shear is 10,400 lb [46.28 kN], the magnitude of the left reaction. On inspecting the shear diagram, we see that the shear passes through zero (the base line) at $x = 10$ ft [3.0 m] directly under the 12,000 lb [53.4 kN] load. Later we show in the design of beams that it is important to know the location of this section. It is here that the greatest bending stresses occur.

So far we have considered a beam with only concentrated loads; now let us make a shear diagram for a uniformly distributed load.

Example 2. The beam shown in Figure 7-6a has a span of 18 ft [5.4 m] and a uniformly distributed load of 500 lb per ft [7.3 kN/m]. Construct the shear diagram by noting the maximum shear and the section at which the shear passes through zero.

Solution: The total load is $500 \times 18 = 9000$ lb [$7.3 \times 5.4 = 39.42$ kN]. Each of the reactions is equal to one-half the total load, or 4500

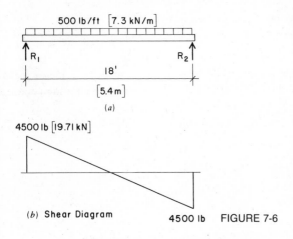

(a)

(b) Shear Diagram   4500 lb   FIGURE 7-6

lb [19.71 kN], which is also the value of the maximum shear at the end of the beam. At a section 1 ft to the right of the left reaction the shear is equal to the reaction minus the increment of load on the 1-ft length of beam. Thus

$$V_{(x=1)} = 4500 - (500 \times 1) = 4000 \text{ lb}$$

$$[V_{(x=0.3)} = 19.71 - (7.3 \times 0.3) = 17.52 \text{ kN}]$$

The shears at some additional sections are as follows:

$$V_{(x=2)} = 4500 - (500 \times 2) = 3500 \text{ lb}$$

$$V_{(x=3)} = 4500 - (500 \times 3) = 3000 \text{ lb}$$

$$V_{(x=9)} = 4500 - (500 \times 9) = 0$$

$$V_{(x=12)} = 4500 - (500 \times 12) = -1500 \text{ lb}$$

$$V_{(x=18)} = 4500 - (500 \times 18) = -4500 \text{ lb}$$

$$
\begin{bmatrix}
V_{(x=0.6)} = 19.71 - (7.3 \times 0.6) = 15.33 \text{ kN} \\
V_{(x=0.9)} = 19.71 - (7.3 \times 0.9) = 13.14 \text{ kN} \\
V_{(x=2.7)} = 19.71 - (7.3 \times 2.7) = 0 \\
V_{(x=3.6)} = 19.71 - (7.3 \times 3.6) = -6.57 \text{ kN} \\
V_{(x=5.4)} = 19.71 - (7.3 \times 5.4) = -19.71 \text{ kN}
\end{bmatrix}
$$

A plotting of these points, as shown in Figure 7-6*b*, indicates that the shear diagram consists of a sloping straight line that passes through zero at the center of the span.

Example 3.   Draw the shear diagram and determine the value of the maximum shear and the section at which the shear passes through zero for the beam shown in Figure 7-7*a*.

Solution: To compute the reactions first take R_2 as the center of moments. Thus

$$16R_1 = (6000 \times 10) + (200 \times 16 \times 8)$$

$$16R_1 = 85,600 \quad \text{and} \quad R_1 = 5350 \text{ lb}$$

$$
\begin{bmatrix}
4.8R_1 = (26.7 \times 3.0) + (2.92 \times 4.8 \times 2.4) \\
4.8R_1 = 113.7384 \quad \text{and} \quad R_1 = 23.70 \text{ kN}
\end{bmatrix}
$$

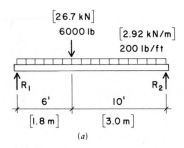

(a)

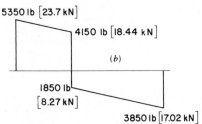

(b)

3850 lb [17.02 kN]   FIGURE 7-7

With R_1 as the center of moments,

$$16R_2 = (6000 \times 6) + (200 \times 16 \times 8) =$$

$$16R_2 = 61{,}600 \quad \text{and} \quad R_2 = 3850 \text{ lb}$$

$$\left[\begin{array}{l} 4.8R_2 = (26.7 \times 1.8) + (2.92 \times 4.8 \times 2.4) \\ 4.8R_2 = 81.6984 \quad \text{and} \quad R_2 = 17.02 \text{ kN} \end{array} \right]$$

We know that the shear at R_1 is 5350 lb [23.70 kN]. Now let us compute the shear at a section infinitely close to, and to the left of, the concentrated load. We call this distance $(x = 6-)$ from R_1. Then

$$V_{(x=6-)} = 5350 - (200 \times 6) = 4150 \text{ lb}$$

$$[V_{(x=1.8-)} = 23.70 - (2.92 \times 1.8) = 18.44 \text{ kN}]$$

We next consider the section just to the right of the concentrated load; call it $(x = 6+)$ from R_1. Then

$$V_{(x=6+)} = 5350 - (200 \times 6) - 6000 = -1850 \text{ lb}$$

$$[V_{(x=1.8+)} = 23.70 - (2.92 \times 1.8) - 26.7 = -8.27 \text{ kN}]$$

Using these two computed values and the known values of shear at the

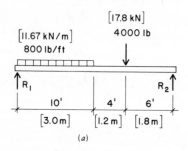

(a)

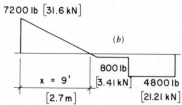

FIGURE 7-8

ends of the beam, we can construct the diagram, as shown in Figure 7-7b, noting that the maximum shear is at R_1 and that the shear passes through zero under the concentrated load.

Example 4.   Construct the shear diagram for the beam shown in Figure 7-8a.

Solution: With R_2 as the center of moments,

$$20R_1 = (800 \times 10 \times 15) + (4000 \times 6)$$

$$20R_1 = 144{,}000 \quad \text{and} \quad R_1 = 7200 \text{ lb}$$

$$\left[\begin{array}{l} 6.0R_1 = (11.67 \times 3.0 \times 4.5) + (17.8 + 1.8) \\ 6.0R_1 = 189.585 \quad \text{and} \quad R_1 = 31.60 \text{ kN} \end{array} \right]$$

With R_2 as the center of moments,

$$20R_2 = (800 \times 10 \times 5) + (4000 \times 14)$$

$$20R_2 = 96{,}000 \quad \text{and} \quad R_2 = 4800 \text{ lb}$$

$$\left[\begin{array}{l} 6.0R_2 = (11.67 \times 3.0 \times 1.5) + (17.8 \times 4.2) \\ 6.0R_2 = 127.275 \quad \text{and} \quad R_2 = 21.21 \text{ kN} \end{array} \right]$$

We observe that the value of the shear at the left end is the same as R_1 and that at the right end is the same as R_2. Computing the shears at other critical points, we obtain

$$V_{(x=10)} = 7200 - (800 \times 10) = -800 \text{ lb}$$

$$V_{(x=14-)} = 7200 - (800 \times 10) = -800 \text{ lb}$$

$$V_{(x=14+)} = 7200 - (800 \times 10) - 4000 = -4800 \text{ lb}$$

$$\left[\begin{array}{l} V_{(x=3.0)} = 31.60 - (11.67 \times 3.0) = -3.41 \text{ kN} \\ V_{(x=4.2-)} = 31.60 - (11.67 \times 3.0) = -3.41 \text{ kN} \\ V_{(x=4.2+)} = 31.60 - (11.67 \times 3.0) - 17.8 = -21.21 \text{ kN} \end{array} \right]$$

These various points are plotted; for the shear diagram see Figure 7-8*b*.

An inspection of the shear diagram shows that the shear passes through zero at some point between R_1 and the end of the distributed load. We call this distance x and write an equation for the shear at this distance. Thus

$$V = 0 = 7200 - (800 \times x)$$

$$800x = 7200 \quad \text{and} \quad x \stackrel{*}{=} 9 \text{ ft}$$

$$\left[\begin{array}{l} 0 = 31.60 - (11.67 \times x) \\ x = 31.60/11.67 = 2.7 \text{ m} \end{array} \right]$$

Example 5. Construct the shear diagram for the beam shown in Figure 7-9*a*.

Solution: With R_2 as the center of moments,

$$20R_1 + (6000 \times 4) = (4000 \times 26) + (8000 \times 14)$$

$$20R_1 = 192,000 \quad \text{and} \quad R_1 = 9600 \text{ lb}$$

$$\left[\begin{array}{l} 6.0R_1 + (26.7 \times 1.2) = (17.8 \times 7.8) + (35.6 \times 4.2) \\ 6.0R_1 = 256.32 \quad \text{and} \quad R_1 = 42.72 \text{ kN} \end{array} \right]$$

With R_1 as the center of moments,

$$20R_2 + (4000 \times 6) = (8000 \times 6) + (6000 \times 24)$$

$$20R_2 = 168,000 \quad \text{and} \quad R_2 = 8400 \text{ lb}$$

$$\left[\begin{array}{l} 6.0R_2 + (17.8 \times 1.8) = (35.6 \times 1.8) + (26.7 \times 7.2) \\ 6.0R_2 = 224.28 \quad \text{and} \quad R_2 = 37.38 \text{ kN} \end{array} \right]$$

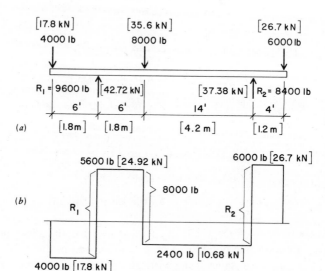

FIGURE 7-9

Then

$$V_{(x=1)} = -4000 \text{ lb}$$

$$V_{(x=7)} = 9600 - 4000 = 5600 \text{ lb}$$

$$V_{(x=13)} = 9600 - 4000 - 8000 = -2400 \text{ lb}$$

$$V_{(x=27)} = 9600 + 8400 - 4000 - 8000 = 6000 \text{ lb}$$

$$\left[\begin{array}{c} V_{(x=0.3)} = -17.8 \text{ kN} \\ V_{(x=2.1)} = 42.72 - 17.8 = 24.92 \text{ kN} \\ V_{(x=3.9)} = 42.72 - 17.8 - 35.6 = -10.68 \text{ kN} \\ V_{(x=8.1)} = 42.72 + 37.38 - 17.8 - 35.6 = 26.7 \text{ kN} \end{array} \right]$$

The shear at other sections might be computed, but we have sufficient information for our purpose. The values computed are now plotted with respect to the base line, and the shear diagram shown in Figure 7-9b results.

This diagram differs somewhat from those previously constructed. For simple beams the maximum shear is the value of the greater reaction; but this is an overhanging beam, and we see that the maximum

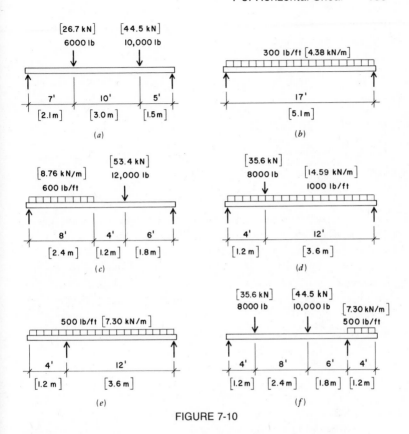

FIGURE 7-10

shear has a value of 6000 lb [26.7 kN]. We observe also that in this beam the shear passes through zero at three different points; at the two supports and under the 8000 lb [35.6 kN] load. The significance is explained later.

Problems 7.4.A-B-C-D*-E*-F.   Construct the shear diagrams for the beams shown in Figure 7-10a, b, c, d, e, and f. In each case note the magnitude of the maximum shear and the section at which the shear passes through zero.

7-5.   Horizontal Shear

Figure 7-11a represents a number of boards placed flat, one above the other, and supported at the ends. A load placed on the boards, or even the weight of the boards alone, tends to cause slipping along the sur-

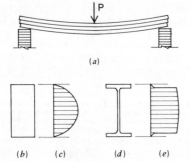

FIGURE 7–11. Horizontal shear in beams.

faces of contact between adjacent boards, as indicated in the figure. This same tendency is present in a solid beam but the action is restrained by its resistance to *horizontal shear*.

It can be shown that at any point in a beam the intensity of the horizontal shear is equal to the intensity of the vertical shear. The shearing stresses, however, are not uniformly distributed over the cross-sectional area of the beam. The unit shearing stress may be found by the formula

$$v = \frac{VQ}{Ib}$$

where

v = the unit horizontal shearing stress at any specific point in the cross section of a beam in pounds per square inch

V = the total vertical shear in the beam at the section selected in pounds

Q = the statical moment with respect to the neutral axis of the area of the cross section above (or below) the point at which v is to be determined (a statical moment is an area multiplied by the distance of its centroid to a given axis) in cubed inches

I = the moment of inertia of the cross section of the beam with respect to its neutral axis in inches to the fourth power

b = the width of the beam at the point at which v is to be computed in inches

The maximum unit horizontal shearing stress in a rectangular beam occurs at the neutral surface, and its magnitude is given by the equation

$$v = \frac{3}{2} \times \frac{V}{bd}$$

This may be demonstrated by considering the rectangular cross section of width b and depth d shown in Figure 7-11b. Let us determine the unit horizontal shearing stress at the neutral axis of the section. The area above the neutral axis is $(b + d/2)$, and its centroid is $d/4$ distance from the neutral axis $X - X$. Therefore

$$Q = b \times \frac{d}{2} \times \frac{d}{4} = \frac{bd^2}{8}$$

I, the moment of inertia of the cross section is equal to $bd^3/12$, Art. 6-3. Then

$$v = \frac{VQ}{Ib} = \frac{V \times (bd^2/8)}{(bd^3/12) \times b} \quad \text{and} \quad v = \frac{3}{2} \times \frac{V}{bd}$$

It should be remembered that the total horizontal shear over the cross-sectional area is V; the maximum unit stress occurs at the neutral surface, and this is the stress with which we are principally concerned. Figure 7-11c indicates the distribution of horizontal shearing stresses in a rectangular cross section. Since in solid timber beams there is less resistance to shear parallel to the grain than perpendicular to it, the design of a timber beam should include investigation for horizontal shear.

From the foregoing discussion it is apparent that a loaded timber beam may fail by horizontal shear as well as by bending. Both of these stresses should be investigated. The unit horizontal shearing stress should always be computed for beams having short spans with large loads. To find the maximum unit horizontal shearing stress in a loaded beam, we consider the section at which V has its maximum value.

Example.   A 10 × 14-in. Douglas Fir timber of Select Structure grade is used for a beam having a span of 16 ft [4.8 m]. There is a uniformly distributed load (including the beam weight) of 400 lb/ft [5.84 kN/m] extending over the entire span and, in addition, a concentrated load of 4000 lb [17.8 kN] is located 4 ft [1.2 m] from the left end. Compute the maximum unit horizontal shearing stress. Is the beam safe with respect to horizontal shear?

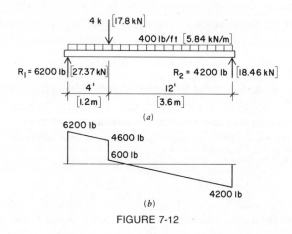

FIGURE 7-12

Solution: (1) The beam diagram, Figure 7-12*a*, is drawn and the reactions computed.

$$16R_1 = (4000 \times 12) + (400 \times 16 \times 8)$$

$$\text{and } R_1 = 6200 \text{ lb } [27.37 \text{ kN}]$$

$$16R_2 = (4000 \times 4) + (400 \times 16 \times 8)$$

$$\text{and } R_2 = 4200 \text{ lb } [18.46 \text{ kN}]$$

(2) Next, the shear diagram is drawn as shown in Figure 7-12*b*. We note that V, the maximum vertical shear, has a magnitude of 6200 lb immediately to the right of the left reaction. Since the actual dimensions of a nominal 10 × 14 in. timber are 9.5 × 13.5 in (Table 6-7),

$$v = \frac{3}{2} \times \frac{V}{bd} = 1.5 \times \frac{6200}{9.5 \times 13.5} = 72.5 \text{ psi } [503 \text{ kPa}]$$

(3) This value must be checked against the allowable unit horizontal shearing stress for Douglas Fir of the grade and size classification specified. Referring to Table 5-3, we find that for Douglas Fir, Select Structural grade, "beams and stringer" classification, the allowable stress F_v is 85 psi. Since the actual stress is less than the allowable, the beam is safe with respect to horizontal shear. Note that the symbol F_v is used in Table 5-3 instead of the letter v. Notation is not entirely

consistent in structural design usage among various references. Reference to Table 6-7 will show that dimension h used therein corresponds to d in the formula for unit shear stress.

Problem 7-5-A.   A 6 × 12 Southern Pine beam, No. 1 SR grade, has a span of 16 ft [4.8 m]. It supports a uniformly distributed load of 400 lb/ft [5.84 kN/m] including its own dead weight. Compute the maximum unit horizontal shearing stress. Is the beam safe with respect to shear?

Problem 7-5-B*.   A 10 × 12 beam has a span of 10 ft [3.0 m]. It carries a uniformly distributed load of 100 lb/ft [1.5 kN/m] including its own dead weight. In addition, a concentrated load of 6000 lb [27 kN] is applied 5 ft [1.5 m] from the left reaction, and another concentrated load of 4000 lb [18 kN] occurs 2 ft [0.6 m] from the right reaction. If the timber is Douglas Fir, Select Structural grade, is the beam safe with respect to horizontal shear?

Problem 7-5-C.   A 3 × 10 cantilever beam projects 6 ft [1.8 m] from the face of a masonry wall. It carries a uniformly distributed load of 100 lb/ft [1.5 kN/m], including its own weight, over its entire length and a concentrated load of 400 lb [1.8 kN] at the free (unsupported) end. The timber is Douglas Fir No. 1 grade. Is this size sufficiently large to resist horizontal shear?

7-6.   Shearing Stresses in Steel Beams

Structural steel beams are not rectangular in cross section and consequently the equation

$$v = \frac{3}{2} \times \frac{V}{bd} = \frac{3}{2} \times \frac{V}{A}$$

must be modified when investigating wide flange shapes and I-beams where most of the material lies in the flanges. Even in beams with these cross sections, however, the maximum shearing stress occurs at the neutral surface and the value at the extreme fibers is zero, so the flanges have little influence on resistance to shear. It is, therefore, customary to ignore the material in the flanges and to consider only the web as resisting the shear. Based on this assumption, the following approximate expression for unit shearing stress is the one customarily employed:

$$f_v = \frac{V}{A_w} = \frac{V}{dt_w}$$

in which

f_v = actual unit shearing stress is psi or ksi
V = maximum vertical shear in pounds of kips
A_w = gross area of the web in square inches
d = overall depth of the beam in inches
t_w = thickness of the beam web in inches

Here, of course, the value of f_v is really the *average* unit shearing stress over the area of the web. The fact that the maximum unit stress is somewhat greater than the average value is handled in practice by assigning a value for the *allowable* shearing stress low enough to compensate for this difference. For A36 steel this value is 14,500 psi or 14.5 ksi [100 MPa], as given on the line with the symbol F_v in Table 5-2.

Example. A W 12 × 40 is used as a beam subjected to a vertical shear at the supports of 50 kips [222 kN]. Find the value of the unit shearing stress.

Solution: Referring to Table 6-1, we find that the depth of this section is 11.94 in. [303 mm] and the web thickness is 0.295 in. [7.49 mm]. Then the area of the web is 11.94 × 0.295 = 3.52 in.^{2} [2270 mm^2]. Therefore

$$f_v = \frac{V}{A_w} = \frac{50}{3.52} = 14.2 \text{ ksi } [97.8 \text{ MPa}]$$

Problem 7-6-A. If the maximum vertical shear force on an S 18 × 70 is 100 kips [445 kN], determine the value of the unit shearing stress.

Problem 7-6-B*. If the maximum vertical shear force on a W 18 × 76 is 100 kips [445 kN], determine the value of the unit shearing stress.

8

Bending Moments in
Beams
IIIIIIIIIIIIIIIIIIIIIIIIIIIIIIIIIIIIIIIIIIIIII

8-1. Introduction

In Chapter 4 we learned that the *moment of a force* is the tendency of
a force to cause rotation about a certain point or axis. The point or axis
is called the *center of moments*, and the perpendicular distance between
the line of action of the force and the point or axis is the *lever arm* or
moment arm. The magnitude of the moment is the magnitude of the
force (pounds, kips, etc.) multiplied by the lever arm (feet, inches,
etc.). Moments are expressed in compound units such as foot-pounds
and inch-pounds, or kip-feet and kip-inches. In computations involving
moments take particular care to designate the units properly; this will
aid in avoiding errors. If a force tends to produce clockwise rotation,
it is customary to designate the moment as *positive* (+20,000 ft-lb,
+20 kip-ft, etc.). Similarly, if the tendency to rotate is counterclockwise, the moment is *negative* (−30 kip-ft, −360 kip-in, etc.). This
sign convention is frequently modified, however, when dealing with
bending moments in beams.

8-2. Bending Moments

As noted earilier, a beam deforms by bending under the action of applied loads. Figure 7-1*a* shows the deformation curve for a simple beam

145

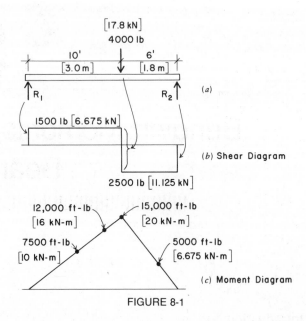

FIGURE 8-1

that supports a concentrated load at midspan; the deformation curve for the beam in Figure 8-1a would have the same general form. At any point between the left reaction and the 4000-lb [17.8-kN] load the tendency for the beam to bend is measured by the moment of the left reaction about the point in question; for example, consider a point 5 ft [1.5 m] to the right of the left support. The moment of the reaction about this point as the center of moments is 1500 × 5 or 7500 ft-lb [6.675 × 1.5 = 10 kN-m]. This moment is called the *bending moment*, and its value may be plotted as a point on a moment diagram (Figure 8-1c). The magnitude of the bending moment varies at different sections along the beam span. Thus

$$M_{(x=5)} = (1500 \times 5) = 7500 \text{ ft-lb}$$

$$M_{(x=8)} = (1500 \times 8) = 12{,}000 \text{ ft-lb}$$

$$M_{(x=10)} = (1500 \times 10) = 15{,}000 \text{ ft-lb}$$

$$\left[\begin{array}{l} M_{(x=1.5)} = (6.675 \times 1.5) = 10 \text{ kN-m} \\ M_{(x=2.4)} = (6.675 \times 2.4) = 16 \text{ kN-m} \\ M_{(x=3.0)} = (6.675 \times 3.0) = 20 \text{ kN-m} \end{array} \right]$$

Suppose we use the same procedure but take the reference from the other reaction. Then

$$M_{(x=2)} = (2500 \times 2) = 5000 \text{ ft-lb}$$

$$M_{(x=6)} = (2500 \times 6) = 15,000 \text{ ft-lb}$$

$$\left[\begin{array}{l} M_{(x=0.6)} = (11.125 \times 0.6) = 6.675 \text{ kN-m} \\ M_{(x=1.8)} = (11.125 \times 1.8) = 20 \text{ kN-m} \end{array} \right]$$

The last value computed is the same as that obtained by proceeding from the other end. Thus we observe the following:

The bending moment at any section of a beam is equal to the algebraic sum of the moments of all forces on one side of the section.

For convenience call upward forces (the reactions) positive quantities, the downward forces (the loads) negative, and consider the forces to the left of the section. Then, in conformity with the foregoing statement, we can make the following rule:

The bending moment at any section of a beam is equal to the moments of the reactions minus the moments of the loads to the left of the section.

Actually this rule will also apply if the reactions and loads to the right of the section are used, as demonstrated in the previous computations. As we have already established the practice of proceeding from the left in the construction of the shear diagram, it will prove to be significant to do the same with moments.

8-3. Bending Moment Diagrams

Fig. 8-1c is a bending moment diagram, four ordinates of which were computed in the preceding section. Bending moment diagrams are constructed quite like the shear diagrams. A base line ($M = 0$) is drawn, and the values of the bending moments at various sections along the beam are plotted to scale. We compute the magnitudes of the moments in accordance with the rule given in Section 8-2. The section at which the moment is to be computed is noted by using the subscripts $x = 1$, $x = 2$, and so on. Then we write the moments of the reaction and subtract the moments of the loads to the left of the section. When we have established a sufficient number of values of bending moments we may construct the complete diagram.

In the design of any beam we are particularly concerned with the maximum value of the bending moment. It will be noted that this value always occurs at a point at which the shear diagram passes through zero, which is why we have emphasized noting these points when shear diagrams are constructed.

8-4. Positive and Negative Bending Moments

Figure 8-2a illustrates the shape a simple beam tends to assume when it bends. The fibers in the upper surface of the beam are in compression; those in the lower surface are in tension. We say the beam is "concave upward" and define the bending moment under this condition as *positive*. Now refer to Figure 8-2b, an overhanging beam. From the sketch we see that a portion of the beam in the vicinity of the right reaction has tension in the upper fibers and compression in the lower; here the beam is "concave downward," and the bending moment is negative in sign. The section at which the bending moment changes from positive to negative is called the *inflection point*. As the following problem demonstrates, the inflection point corresponds to a value of zero moment on the moment diagram.

Problem 8.4.A.   Construct the bending moment diagram for the beam shown in Fig. 7-9a and note the value of the maximum bending moment.

8-5. Concentrated Load at Center of Span

A condition that occurs frequently in practice is a concentrated load at the center of the span of a simple beam. In Figure 8-3a let P denote

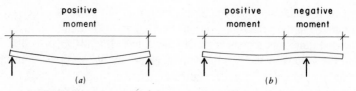

FIGURE 8-2.   Sign of bending moment: bending stress convention.

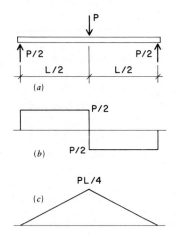

FIGURE 8-3

the concentrated load and L the span length. Then, as the beam is symmetrical, each reaction is equal to $P/2$; see the shear diagram in Figure 8-3b. This diagram shows that the shear passes through zero at the center of the span at $x = L/2$. At this section the value of the bending moment will be maximum. By following the rule given in Section 8-2 its magnitude is readily found:

$$M_{(x = L/2)} = \frac{P}{2} \times \frac{L}{2} = \frac{PL}{4} \text{ (See Figure 8-3)}$$

This is a most useful expression; for instance, suppose we are to design a beam with a span of 24 ft [7.2 m] and a concentrated load of 18 kips [80 kN] at midspan. It will be necessary to compute the maximum bending moment, which is readily done for we have found that its value is $PL/4$. Thus

$$M = \frac{PL}{4} = \frac{18 \times 24}{4} = 108 \text{ k-ft}$$

$$\left[\frac{80 \times 7.2}{4} = 144 \text{ kN-m} \right]$$

If the bending moment is required in units of kip-in., as it frequently is, $M = 108 \times 12 = 1296$ k-in.

8-6. Simple Beam with Uniformly Distributed Load

Another condition that occurs perhaps more times than any other is represented by a simple beam with a uniformly distributed load (Figure 8-4a). To develop a formula for its maximum bending moment let L be the span in feet and w the uniform load in pounds per linear foot. Then the total load is wL, each reaction is $wL/2$, and the shear diagram takes the form shown in Figure 8-4b. (See also Figure 7-6.) The maximum bending moment is at the center of the span. Then by application of the rule given in Section 8-2 the value of the bending moment at midspan is

$$M = \left(\frac{wL}{2} \times \frac{L}{2}\right) - \left(\frac{wL}{2} \times \frac{L}{4}\right) = \frac{wL^2}{8}$$

In this expression the load per linear foot is w lb, and the total load is wL. If the total load is represented by W, we have

$$W = wL \text{ and } M = \frac{wL^2}{8} = \frac{WL}{8}$$

which is another useful form of the equation. If moments are computed for other sections along the beam span, the values will plot to form a parabola with its apex at midspan (Figure 8-4c).

Example. A simple beam with a span of 22 ft [6.6 m] has a uniformly distributed load of 800 lb/ft [12 kN/m]. Compute the maximum bending moment.

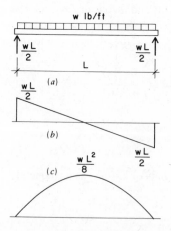

FIGURE 8-4

Solution: Substituting in the formula that was previously derived, we obtain

$$M = \frac{wL^2}{8} = \frac{800 \times 22 \times 22}{8} = 48,400 \text{ ft-lb}$$

$$\left[M = \frac{12 \times 6.6 \times 6.6}{8} = 65.34 \text{ kN-m} \right]$$

This is the value of the bending moment at the center of the span. It is computed quickly by using the formula. If, however, we were required to construct the entire bending moment diagram, it would be necessary to compute the values at several other sections so that a smooth curve could be plotted. In this example R_1 is equal to half the total load on the beam, or 8800 lb [39.6 kN]; the bending moments at some other points are as follows:

$$M_{(x=4)} = (8800 \times 4) - (800 \times 4 \times 2) = 28,800 \text{ ft-lb}$$

$$M_{(x=8)} = (8800 \times 8) - (800 \times 8 \times 4) = 44,800 \text{ ft-lb}$$

$$M_{(x=11)} = (8800 \times 11) - (800 \times 11 \times 5.5) = 48,400 \text{ ft-lb}$$

$$M_{(x=16)} = (8800 \times 16) - (800 \times 16 \times 8) = 38,400 \text{ ft-lb}$$

$$\left[\begin{array}{l} M_{(x=1.2)} = (39.6 \times 1.2) - (12 \times 1.2 \times 0.6) = 38.88 \text{ kN-m} \\ M_{(x=2.4)} = (39.6 \times 2.4) - (12 \times 2.4 \times 1.2) = 60.48 \text{ kN-m} \\ M_{(x=3.3)} = (39.6 \times 3.3) - (12 \times 3.3 \times 1.65) = 65.34 \text{ kN-m} \\ M_{(x=4.8)} = (39.6 \times 4.8) - (12 \times 4.8 \times 2.4) = 51.84 \text{ kN-m} \end{array} \right]$$

The reader should verify that a plot of these points will form a moment diagram like that in Figure 8-4c.

Problem 8.4-A.   A simple beam has a span of 16 ft [4.8 m] and a uniformly distributed load of 600 lb/ft [8.76 kN/m]. Compute the bending moment at the center and at the quarter points of the span and draw the shear and bending moment diagrams.

8-7. Maximum Bending Moments

To compute the magnitude of the maximum bending moment for a beam we must first know its position along the span. For an unsymmetrical loading it is generally necessary to construct the shear diagram. Then, as noted earlier, the section at which the shear passes

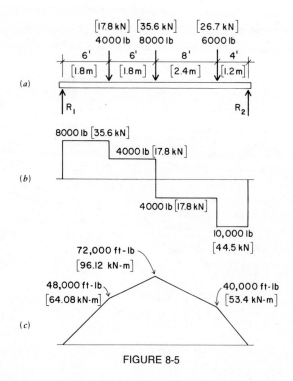

FIGURE 8-5

through zero is the section at which maximum bending moment occurs. The following examples relate to unsymmetrically loaded beams and show the method of determining the maximum bending moments.

Example 1.   The simple beam shown in Figure 8-5a has three concentrated loads at the positions indicated. Compute the maximum bending moment.

Solution: To find the reactions we may determine R_1 by writing a moment equation and R_2 by summing the vertical forces.

$$24R_1 = (4000 \times 18) + (8000 \times 12)$$
$$+ (6000 \times 4)$$
$$24R_1 = 192,000 \quad \text{and} \quad R_1 = 8000 \text{ lb}$$
$$4000 + 8000 + 6000 = 8000 + R_2$$
$$18,000 - 8000 = R_2 \quad \text{and} \quad R_2 = 10,000 \text{ lb}$$

$$\left[\begin{array}{l} 7.2R_1 = (17.8 \times 5.4) + (35.6 \times 3.6) + (26.7 \times 1.2) \\ \qquad 7.2R_1 = 256.32 \quad \text{and} \quad R_1 = 35.6 \text{ kN} \\ \qquad 17.8 + 35.6 + 26.7 = 35.6 + R_2 \\ \qquad 80.1 - 35.6 = R_2 \quad \text{and} \quad R_2 = 44.5 \text{ kN} \end{array}\right]$$

The value of the shear is computed at critical sections; the shear diagram is constructed as shown in Figure 8-5b. In this diagram we find that the shear passes through zero under the 8000-lb [35.6-kN] load. The maximum bending moment is therefore

$$M_{(x=12)} = (8000 \times 12) - (4000 \times 6) = 72,000 \text{ ft-lb}$$

$$[M_{(x=3.6)} = (35.6 \times 3.6) - (17.8 \times 1.8) = 96.12 \text{ kN-m}]$$

To construct the bending moment diagram the expressions for moments under the other loads are

$$M_{(x=6)} = 8000 \times 6 = 48,000 \text{ ft-lb}$$

$$M_{(x=20)} = (8000 \times 20) - (4000 \times 14) - (8000 \times 8)$$

$$= 40,000 \text{ ft-lb}$$

$$\left[\begin{array}{l} \qquad M_{(x=1.8)} = 35.6 \times 1.8 = 64.08 \text{ kN-m} \\ M_{(x=6.0)} = (35.6 \times 6.0) - (17.8 \times 4.2) - (35.6 \times 2.4) \\ \qquad\qquad = 53.40 \text{ kN-m} \end{array}\right]$$

The diagram in Figure 8-5c may now be constructed.

Example 2. A simple beam with a span of 16 ft [4.8 m] sustains the loading shown in Figure 8-6a. Compute the maximum bending moment.

Solution: Determine R_1 and R_2 by using one moment equation and then summing the vertical forces:

$$16R_1 = (400 \times 16 \times 8) + (10,000 \times 6)$$

$$16R_1 = 111,200 \quad \text{and} \quad R_1 = 6950 \text{ lb}$$

$$(400 \times 16) + 10,000 = 6950 + R_2$$

$$16,400 - 6950 = R_2 \quad \text{and} \quad R_2 = 9450 \text{ lb}$$

$$\begin{bmatrix} 4.8R_1 = (5.84 \times 4.8 \times 2.4) + (44.5 \times 1.8) \\ 4.8R_1 = 147.3768 \quad \text{and} \quad R_1 = 30.70 \text{ kN} \\ (5.84 \times 4.8) + 44.5 = 30.70 + R_2 \\ 72.532 - 30.70 = R_2 \quad \text{and} \quad R_2 = 41.832 \text{ kN} \end{bmatrix}$$

In constructing the shear diagram (Figure 8-6b), we find that the maximum bending moment (point of zero shear) occurs under the concentrated load. Thus

$$M_{(x=10)} = (6950 \times 10) - (400 \times 10 \times 5) = 49,500 \text{ ft-lb}$$

$$[M_{(x=3.0)} = (30.7 \times 3.0) - (5.84 \times 3.0 \times 1.5) = 65.82 \text{ kN-m}]$$

To draw the bending moment diagram more accurately, two additional values may be computed:

$$M_{(x=6)} = (6950 \times 6) - (400 \times 6 \times 3) = 34,500 \text{ ft-lb}$$

$$M_{(x=14)} = (6950 \times 14) - (400 \times 14 \times 7) - (10,000 \times 4)$$

$$= 18,100 \text{ ft-lb}$$

$$\begin{bmatrix} M_{(x=1.8)} = (30.7 \times 1.8) - (5.84 \times 1.8 \times 0.9) = 45.80 \text{ kN-m} \\ M_{(x=4.2)} = (30.7 \times 4.2) - (5.84 \times 4.2 \times 2.1) - (44.5 \times 1.2) \\ = 24.03 \text{ kN-m} \end{bmatrix}$$

Example 3. Compute the maximum bending moment for the beam shown in Figure 7-8a.

Solution: Inspection of the loading diagram (Figure 7-8a) gives no indication of the position of the maximum bending moment. However, after computing the reactions and drawing the shear diagram (Figure 7-8b) it was found that the shear passed through zero at a section 9 ft [2.7 m] from the left support. The value of the bending moment at this section is

$$M_{(x=9)} = (7200 \times 9) - (800 \times 9 \times 4.5) = 32,400 \text{ ft-lb}$$

$$\begin{bmatrix} M_{(x=2.7)} = (31.6 \times 2.7) - (11.67 \times 2.7 \times 1.35) \\ = 42.78 \text{ kN-m} \end{bmatrix}$$

8-8.   Overhanging Beams

For the simple beams previously discussed we have seen that the maximum shear has the same magnitude as the greater reaction. This is not
true for overhanging beams; for example, we found that the beam
shown in Figure 7-9, Section 7-4, had R_1 as the greater reaction, with
a value of 9600 lb [42.72 kN]. The maximum shear, on the other hand,
was 6000 lb [26.7 kN] and occurred just to the right of R_2. We noted
also that the shear diagram for this beam passed through zero at three
sections along its length; therefore it is necessary to determine at which
of these sections the actual maximum bending moment occurs. This
determination is accomplished by constructing the shear and moment
diagrams, exercising particular care when plotting the positive and
negative values. We take as an example the beam shown in Figure 7-
9 but restated here as Figure 8-7.

Example.   Construct the shear and bending moment diagrams for the

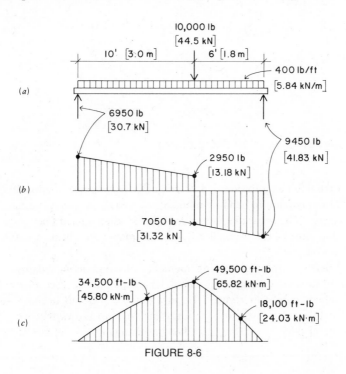

FIGURE 8-6

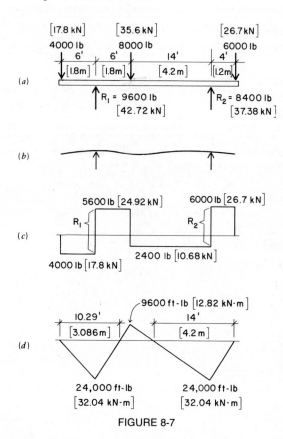

FIGURE 8-7

overhanging beam in Figure 8-7a and note the values of the maximum shear and maximum bending moment.

Solution: The shear diagram is constructed as explained for Figure 7-9b and shown here as Figure 8-7c. The maximum shear is 6000 lb [26.7 kN].

When determining bending moments in overhanging beams it is helpful to sketch the approximate deformation curves. (See Figure 8-7b and compare with Figure 7-1c, in which the overhang occurs at one end only.) The curve for this beam and loading shows that we will encounter positive and negative bending moments, as defined in Section 8-4. Now let us compute the bending moments at certain sections:

$$M_{(x=6)} = -(4000 \times 6) = -24,000 \text{ ft-lb}$$

$$M_{(x=12)} = (9600 \times 6) - (4000 \times 12) = +9600 \text{ ft-lb}$$

$$M_{(x=26)} = (9600 \times 20) - (4000 \times 26) - (8000 \times 14)$$

$$= -24,000 \text{ ft-lb}$$

$$\left[\begin{array}{c} M_{(x=1.8)} = -(17.8 \times 1.8) = -32.04 \text{ kN-m} \\ M_{(x=3.6)} = (42.72 \times 1.8) - (17.8 \times 3.6) = +12.82 \text{ kN-m} \\ M_{(x=7.8)} = (42.72 \times 6.0) - (17.8 \times 7.8) - (35.6 \times 4.2) \\ = -32.04 \text{ kN-m} \end{array} \right]$$

Other values might be computed, but because there are no uniformly distributed loads it is unnecessary. The values just computed are plotted with respect to the baseline: positive values above and negative values below. Figure 8-7d is the completed diagram. The points at which the bending moment diagram passes through zero are of special significance; they are discussed in the following section, and the determination of their location is illustrated. The maximum value of the bending moment is 24,000 ft-lb [32.04 kN-m]; it is negative and occurs over both supports. (It is only coincidental that the moment has the same magnitude over both supports in this example.)

For beams with complicated loading patterns it is advantageous to compute bending moments from both ends to simplify the arithmetic. In a procedure that will always give the correct sign for the bending moment the moments of all upward forces are considered as positive and the moments of all downward forces as negative. Using this convention, we compute the bending moment over the right reaction in Fig. 8-7:

$$M = -(6000 \times 4) = -24,000 \text{ ft-lb}$$

$$[M = -(26.7 \times 1.2) = 32.04 \text{ kN-m}]$$

which is the same value as that obtained, working from the left end, for M at $x = 26$ ft [7.8 m].

8-9. Inflection Point

The two points in Figure 8-7d, where the bending moment diagram passes through zero, are called the *inflection points* or *points* of *contraflexure*. The inflection point is the section along the beam length at which the curvatuve of the beam changes and at which the value of the bending moment diagram is zero. On either side of this section the positions of the tension and compression stresses in the beam section reverse. This may be observed by inspecting the deformed shape (Figure 8-7b).

Example 1.   Determine the position of the inflection point to the right of the left reaction for the beam in Figure 8-7.
Solution: Let x be the distance from the inflection point to the left end of the beam. The expression for bending moment at this point is

$$M = \{9600 \times (x - 6)\} - (4000 \times x)$$

$$[M = \{42.72 \times (x - 1.8)\} - (17.8 \times x)]$$

We know that the value of the moment at this section is zero. Then

$$0 = \{9600 \times (x - 6)\} - (4000 \times x)$$

$$9600x - 57,600 = 4000x$$

$$5600x = 57,600 \quad \text{and} \quad x = 10.29 \text{ ft}$$

$$\left[\begin{array}{c} 0 = \{42.72 \times (x - 1.8)\} - (17.8 \times x) \\ 42.72x - 76.896 = 17.8x \\ 24.92x = 76.896 \quad \text{and} \quad x = 3.086 \text{ m} \end{array} \right]$$

To find the position of the inflection point to the left of the right support it will simplify the mathematics if we consider the moments of the forces to the *right* of the section instead of the left. Let x be the distance from the inflection point to the right end of the beam. Then, as before,

$$0 = \{8400 \times (x - 4)\} - (6000 \times x)$$

$$8400x - 33,600 = 6000x$$

$$2400x = 33,600 \quad \text{and} \quad x = 14 \text{ ft}$$

$$\begin{bmatrix} 0 = \{37.38 \times (x - 1.2)\} - (26.7 \times x) \\ 37.38x - 44.856 = 26.7x \\ 10.68x = 44.856 \quad \text{and} \quad x = 4.2 \text{ m} \end{bmatrix}$$

Example 2. The overhanging beam shown in Figure 8-8*a* has a uniformly distributed load of 200 lb/ft [2.92 kN/m] over its entire length. Construct the shear and moment diagrams, note the values of maximum shear and maximum bending moment, and compute the position of the inflection point.

Solution: The curve the bent beam will take is approximated in Figure 8-8*b*; we expect to find both positive and negative bending mo-

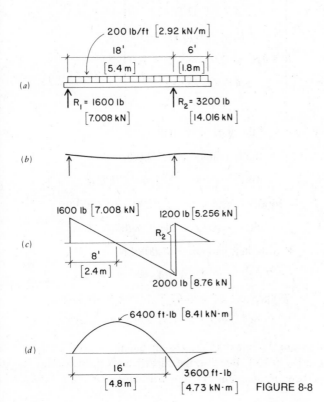

(*a*)

(*b*)

(*c*)

(*d*)

FIGURE 8-8

ments. By computing the reactions we obtain

$$18R_1 = 200 \times 24 \times 6$$

$$18R_1 = 28{,}800$$

$$R_1 = 1600 \text{ lb}$$

$$18R_2 = 200 \times 24 \times 12$$

$$18R_2 = 57{,}600$$

$$R_2 = 3200 \text{ lb}$$

$$\begin{bmatrix} 5.4R_1 = 2.92 \times 7.2 \times 1.8 \\ 5.4R_1 = 37.8432 \\ R_1 = 7.008 \text{ kN} \\ 5.4R_2 = 2.92 \times 7.2 \times 3.6 \\ 5.4R_2 = 75.6864 \\ R_2 = 14.016 \text{ kN} \end{bmatrix}$$

Computing the values of the shears, we have

$$V \text{ at the left support} = 1600 \text{ lb}$$

$$V_{(x=18-)} = 1600 - (200 \times 18) = -2000 \text{ lb}$$

$$V_{(x=18+)} = 1600 - (200 \times 18) + 3200 = +1200 \text{ lb}$$

$$V_{(x=24)} = 1600 - (200 \times 24) + 3200 = 0$$

$$\begin{bmatrix} V \text{ at the left support} = 7.008 \text{ kN} \\ V_{(x=5.4-)} = 7.008 - (2.92 \times 5.4) = -8.760 \text{ kN} \\ V_{(x=5.4+)} = 7.008 - (2.92 \times 5.4) + 14.016 = +5.256 \\ V_{(x=7.2)} = 7.008 - (2.92 \times 7.2) + 14.016 = 0 \end{bmatrix}$$

The shear diagram is plotted in Figure 8-8c, and the maximum shear value is 2000 lb [8.76 kN]. Because the shear passes through zero at two points, the bending moment diagram will have maximum values at two places; one, a positive moment, the other, a negative. The maximum negative moment is directly above the right support. To find the position of zero shear between the supports

$$0 = 1600 - (200 \times x)$$

$$200x = 1600 \quad \text{and} \quad x = 8 \text{ ft}$$

$$\left[\begin{array}{c} 0 = 7.008 - (2.92 \times x) \\ 2.92x = 7.008 \quad \text{and} \quad x = 2.4 \text{ m} \end{array}\right]$$

The bending moment will have values of zero at each end of the beam. Values for the other two critical sections are computed as follows:

$$M_{(x=8)} = (1600 \times 8) - (200 \times 8 \times 4) = 6400 \text{ ft-lb}$$

$$M_{(x=18)} = (1600 \times 18) - (200 \times 18 \times 9) = -3600 \text{ ft-lb}$$

$$\left[\begin{array}{l} M_{(x=2.4)} = (7.008 \times 2.4) - (2.92 \times 2.4 \times 1.2) \\ \qquad = 8.4096 \text{ kN-m} \\ M_{(x=5.4)} = (7.008 \times 5.4) - (2.92 \times 5.4 \times 2.7) \\ \qquad = -4.7304 \text{ kN-m} \end{array}\right]$$

The positive value is the maximum for the beam.

To find the position of the inflection point let x be its distance from the left support. The value of the bending moment at this section is zero. Then

$$(1600 \times x) - \left(200 \times x \times \frac{x}{2}\right) = 0$$

$$\frac{200x^2}{2} - 1600x = 0$$

$$100x^2 - 1600x = 0$$

$$x^2 - 16x = 0$$

$$
\left[
\begin{array}{c}
(7.008 \times x) - \left(2.92 \times x \times \dfrac{x}{2}\right) = 0 \\[2mm]
\dfrac{2.92x^2}{2} - 7.008x = 0 \\[2mm]
1.46x^2 - 7.008x = 0 \\[2mm]
x^2 - 4.8x = 0
\end{array}
\right]
$$

To complete the square

$$x^2 - 16x + 64 = 64$$
$$(x - 8)^2 = 64$$
$$
\left[
\begin{array}{c}
x^2 - 4.8x + 5.76 = 5.76 \\[1mm]
(x - 2.4)^2 = 5.76
\end{array}
\right]
$$

By extracting the square root of both sides we obtain

$$x - 8 = 8 \quad \text{and} \quad x = 16 \text{ ft}$$
$$[x - 2.4 = 2.4 \quad \text{and} \quad x = 4.8 \text{ m}]$$

It may also be observed in this case that the moment diagram will be a simple symmetrical parabola, and the distance to the inflection point will be twice that to the point of zero shear. This will not be true when any concentrated loads fall within this distance.

8-10.   Cantilever Beams

The computations for shear and bending moments for the portion of cantilever beams that extends beyond the face of the support are quite simple. The vertical reaction at the support is equal to the sum of the loads on the beam. If the beam diagram is drawn with the support at the right, the rules given in Sections 7-3 and 8.2 for computing shear and moments are used. The maximum values for both shear and bending moment are at the face of the support.

Example.   The cantilever beam shown in Figure 8-9a has a concentrated load at the free end and a uniformly distributed load applied over

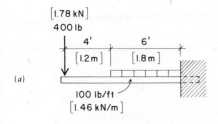

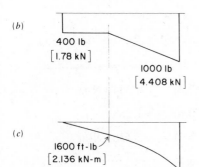

FIGURE 8-9

the first 6 ft [1.8 m] of length from the support. Construct the shear and bending moment diagrams and note the maximum values of each.

Solution: The value of the shear is computed at various sections:

$$V_{(x=1)} = -400 \text{ lb}$$

$$V_{(x=4)} = -400 \text{ lb}$$

$$V_{(x=6)} = -400 - (100 \times 2) = -600 \text{ lb}$$

$$V_{(x=10)} = -400 - (100 \times 6) = -1000 \text{ lb}$$

$$\left[\begin{array}{c} V_{(x=0.3)} = -1.78 \text{ kN} \\ V_{(x=1.2)} = -1.78 \text{ kN} \\ V_{(x=1.8)} = -1.78 - (1.46 \times 0.6) = -2.656 \text{ kN} \\ V_{(x=3.0)} = -1.78 - (1.46 \times 1.8) = -4.408 \text{ kN} \end{array} \right]$$

The maximum value of the shear is at the support; the complete shear diagram is shown in Figure 8-9*b*.

By computing the values of the bending moment at various sections we have

$$M_{(x = 1)} = -(400 \times 1) = -400 \text{ ft-lb}$$

$$M_{(x = 2)} = -(400 \times 2) = -800 \text{ ft-lb}$$

$$M_{(x = 4)} = -(400 \times 4) = -1600 \text{ ft-lb}$$

$$M_{(x = 8)} = -(400 \times 8) - (100 \times 4 \times 2) = -4000 \text{ ft-lb}$$

$$M_{(x = 10)} = -(400 \times 10) - (100 \times 6 \times 3) = -5800 \text{ ft-lb}$$

$$\left[\begin{matrix} M_{(x = 0.3)} = -(1.78 \times 0.3) = -0.534 \text{ kN-m} \\ M_{(x = 0.6)} = -(1.78 \times 0.6) = -1.068 \text{ kN-m} \\ M_{(x = 1.2)} = -(1.78 \times 1.2) = -2.136 \text{ kN-m} \\ M_{(x = 2.4)} = -(.178 \times 2.4) - (1.46 \times 1.2 \times 0.6) = -5.323 \text{ kN-m} \\ M_{(x = 3.0)} = -(1.78 \times 3.0) - (1.46 \times 1.8 \times 0.9) = -7.705 \text{ kN-m} \end{matrix} \right]$$

The maximum bending moment is at the face of the support; its value is -5800 ft-lb [-7.705 kN-m]. The curve of the bending moment is a straight line from the free end up to the beginning of the uniformly distributed load and a portion of a parabola from that point to the support. In accordance with the convention established in Section 8-4 the full length of the cantilever will be under a negative bending moment.

8-11. Typical Loads for Simple and Cantilever Beams

A simple beam with the load at the center of the span and a simple beam with a uniformly distributed load over its entire length are conditions that are frequently met in practice. These and other common types of loading are shown in Figure 8-10. This figure reveals at a glance the values of the maximum shear, maximum bending moment, and maximum deflection (to be discussed later) for the conditions presented. The following notation is used in the figure:

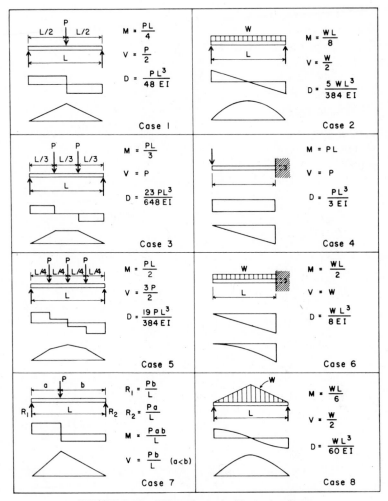

FIGURE 8-10. Values for typical beam loadings.

P = the concentrated load in pounds
W = the total uniformly distributed load in pounds (in Case 8, W represents a triangular loading)
L = the span in feet
V = the maximum shear in pounds
M = the maximum bending moment in foot-pounds
D = the maximum deflection in inches

Problems 8.11.A-B. Construct the shear and bending moment diagrams for the beams shown in Figure 8-11a and b. Note the magnitudes of the maximum shear and the maximum bending moment on the diagrams.

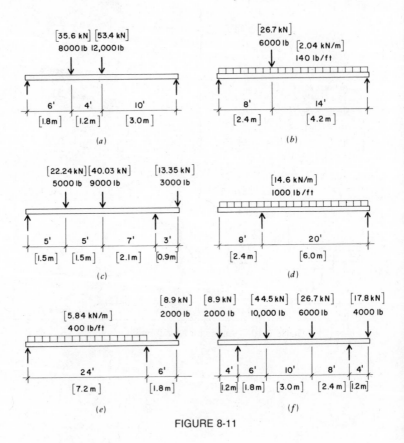

FIGURE 8-11

Problems 8.11.C*-D-E-F. For the beams shown in Figure 8-11c, d, e, and f draw the shear and bending moment diagrams. For each beam give the value of the maximum shear and maximum bending moment and compute the position of the inflection points.

8-12. Bending Moment Determined by Shear Diagram

The method of determining the value of the bending moment as explained previously in this chapter is probably as simple and direct as any. However, the bending moment may be computed by the use of the shear diagram. Since this method is sometimes used, the following explanation is presented.

In shear diagrams the vertical distances represent forces (pounds) and the horizontal distances represent lengths (feet). *The magnitude of the bending moment at any section of a beam is equal to the area of the shear diagram to the left of the section.* Be sure to note the positive and negative values of the shear; in computing the bending moment, it may be necessary to subtract one area from the other.

Example 1. Compute the maximum bending moment for the beam shown in Figure 8-1a.

Solution: The shear diagram (Figure 8-1b) shows that the shear passes through zero at 10 ft [3 m] from the left support. The vertical dimension of the shear diagram to the left of this point is 1500 lb [6.675 kN], and the horizontal length is 10 ft [3 m]. Hence the area of the shear diagram is $1500 \times 10 = 15,000$ ft-lb [20 kN-m]. This is the maximum bending moment.

Example 2. Compute the maximum bending moment for the beam shown in Figure 8-5a.

Solution: The shear diagram (Figure 8-5b) shows that the shear passes through zero 12 ft [3.6 m] from the left support. The area of the shear diagram to the left of this point is composed of two rectangles. Hence the area is

$$(8000 \times 6) + (4000 \times 6) = 72,000 \text{ ft-lb} \quad [96.12 \text{ kN-m}]$$

Example 3. Compute the maximum bending moment for the beam shown in Figure 8-6a.

Solution: In this example the shear diagram comprises a trapezoid. The area may be found as the sum of the areas of a triangle and a rectangle or by finding the area of the trapezoid directly as the product of the base (10 ft) times the average of the heights at the ends (6950 lb and 2950 lb). Thus

$$\frac{(6950 + 2950)}{2} \times 10 = 49,500 \text{ ft-lb} \quad [65.82 \text{ kN-m}]$$

Example 4.   Compute the maximum positive and negative bending moments for the beam shown in Figure 8-8a.

Solution: From the discussion in Example 2 of Section 8-9 it was found that the shear passed through zero at two points, 8 ft [2.4 m] and 18 ft [5.4 m] from the left end of the beam. The value of the maximum positive moment may be found as the area of the triangular portion of the shear diagram extending 8 ft from the left end of the beam. Thus

$$\frac{(1600 \times 8)}{2} = 6400 \text{ ft-lb} \quad [8.41 \text{ kN-m}]$$

The maximum negative moment occurs at R_2. The area of the shear diagram to the left of R_2 consists of two triangles, one negative and one positive. Hence the value of the maximum negative moment is

$$\frac{(+1600 \times 8)}{2} + \frac{(-2000 \times 10)}{2} = +6400 - 10,000$$

$$= -3600 \text{ ft-lb} \quad [-4.73 \text{ kN-m}]$$

The negative moment value at R_2 may also be found by determining the area of the shear diagram to the right of R_2. This is the simpler computation in this case, although if both computations are made they serve to provide a check on the work.

Problems 8-12-A-B-C-D.   Using the shear area method, compute the magnitudes of maximum bending moment for the following beams: (a) Figure 8-3; (b) Figure 8-4; (c) Figure 8-7; (d) Figure 8-9.

Problems 8-12-E-F-G-H-I-J.   Using the shear area method, compute the maximum bending moments for the beams in Figure 8-11, as follows: (e) 8-11a, (f) 8-11b, (g) 8-11c, (h) 8-11d, (i) 8-11e, (j) 8-11f.

9

Continuous and Restrained Beams

IIIIIIIIIIIIIIIIIIIIIIIIIIIIIIIIIIIIIIIIIIII

9-1. Bending Moments for Continuous Beams

It is beyond the scope of this book to give a detailed discussion of bending in members continuous over supports, but the material presented in this chapter will serve as an introduction to the subject. A *continuous beam* was defined in Section 7-1 as a beam that rests on more than two supports. For most continuous beams the maximum bending moment is smaller than that found in a series of simply supported beams having the same spans and loads. Continuous beams are characteristic of reinforced concrete and welded steel floor framing.

The concepts underlying continuity and bending under restraint are illustrated in Figure 9-1. Figure 9-1a represents a single beam resting on three supports and carrying equal loads at the centers of the two spans. If we imagine the beam to be cut over the middle support as shown in Figure 9-1b, the result will be two simple beams. Each of these simple beams will deflect as shown. However, when the beam is made continuous over the support, the deflection curve has a shape similar to that indicated by the dotted line in the first figure.

It is evident that there is no bending moment developed over the support in Figure 9-1b, while there must be a moment over the support

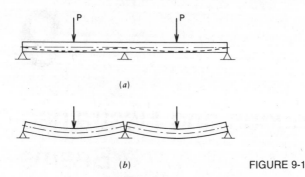

(a)

(b)   FIGURE 9-1

in Figure 9-1a. Study of the figures shows that, in both cases, there is tension in the bottom of the beams near midspan; in addition, there is tension developed in the top of the beam of Figure 9-1a over the center support. In other words, the continuous beam has a positive bending moment near the center of each span and a negative moment over the middle support. It can be shown that the positive bending moment near midspan for the continuous beam is less than that for the simple spans of Figure 9-1b.

The value of the bending moments in continuous beams cannot be found by the usual equations of static equilibrium; additional equations which involve the elasticity of the material are required. Consequently a continuous beam is described as "statically indeterminate." There are several methods of analysis for statically indeterminate structures, and bending moment formulas have been developed for various typical conditions of continuous beam loading and restraint. An extensive table of such formulas is presented under the title "Beam Diagrams and Formulas" in Part 2 of the *Manual of Steel Construction* published by the American Institute of Steel Construction (AISC). These must be used with judgment, however, since the actual conditions under which the structure is built may not duplicate the theoretical ones on which the formulas are based.

Theorem of Three Moments.   One method of determining reactions and constructing the shear and bending moment diagrams for continuous beams is based on the *theorem of three moments*. This theorem deals with the relation among the bending moments at any three consecutive supports. The equation of this relation is known as the *three-moment equation*, and its form varies with the type of loading. When

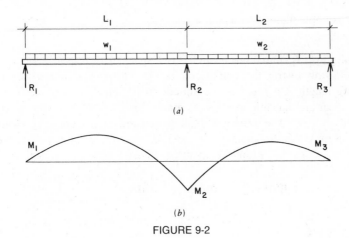

(a)

(b)

FIGURE 9-2

the bending moments at all supports are established, the magnitudes of the reactions may be computed, and, by the methods previously described, the shear and moment diagrams may be drawn.

Figure 9-2a represents a continuous beam of two spans with uniformly distributed loads. If the beam is simply supported at both ends, the bending moment at each of these two supports must be zero, and the bending moment diagram will have the general shape shown in Figure 9-2b. There will be a negative bending moment at the middle support.

The three-moment equation for a continuous beam of two spans with uniformly distributed loads and constant moment of inertia is

$$M_1 L_1 + 2M_2(L_1 + L_2) + M_3 L_2 = -\frac{w_1 L_1^3}{4} - \frac{w_2 L_2^3}{4}$$

in which the various terms are as shown in Figure 9-2 with w expressed in pounds per linear foot. Several simple examples are presented in the following sections to show how the theorem of three moments is applied.

9-2. Continuous Beam with Two Equal Spans

We consider first a continuous beam of two equal spans with the same uniformly distributed load extending the full length of each span. In this example there is no restraint at the end supports (ends simply supported).

Example.   A continuous beam has two spans of 10 ft [3 m] each with a uniformly distributed load of 100 lb/ft [1.5 kN/m] extending over its entire length. Compute the magnitude of the reactions and construct the shear and bending moment diagrams.

Solution: (1) The beam and loading are shown in Figure 9-3a. Bearing in mind that the moments M_1 and M_3 are each zero, substitute the known values in the three-moment equation given before. Then

$$(0 \times 10) + [2M_2(10 + 10)] + (0 \times 10) = -\frac{100 \times 1000}{4} \times 2$$

$$40M_2 = -50,000$$

$$M_2 = -1250 \text{ ft-lb}$$

$$[1.6875 \text{ kN/m}]$$

which is the negative bending moment at the center support.

(2) Next we write an expression for the bending moment at 10 ft from the left support and equate it to the value we have just found (-1250 ft-lb). Then

$$M_{(x=10)} = (R_1 \times 10) - (100 \times 10 \times 5) = -1250$$

$$10R_1 = 3750$$

$$R_1 = 375 \text{ lb } [1.6875 \text{ kN}]$$

Since R_1 and R_3 are equal, R_3 also equals 375 lb.

(3) To find R_2 we note that the total load on the entire length of the beam is $2 \times 100 \times 10 = 2000$ lb, hence

$$R_1 + R_3 + R_2 = 2000$$

$$375 + 375 + R_2 = 2000$$

$$R_2 = 1250 \text{ lb } [5.625 \text{ kN}]$$

(4) Now that the magnitudes of the reactions have been established, we can construct the shear diagram as shown in Figure 9-3b. Let x be the distance in feet from R_1 to the point between R_1 and R_2 at which

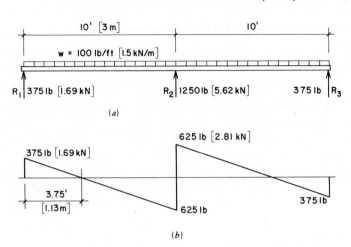

(a)

(b)

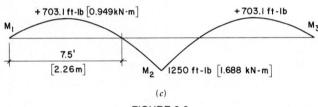

(c)

FIGURE 9-3

the shear diagram passes through zero. Then, writing an expression for the shear at this point and equating it to zero,

$$375 - (100 \times x) = 0 \quad \text{and} \quad x = 3.75 \text{ ft } [1.125 \text{ m}]$$

(5) The maximum positive bending moment is

$$M_{(x=3.75)} = (375 \times 3.75) - \left(3.75 \times 100 \times \frac{3.75}{2}\right)$$

$$= 703.1 \text{ ft-lb } [0.9492 \text{ kN-m}]$$

Since this beam is symmetrical, the positive bending moment between R_2 and R_3 will likewise have a magnitude of 703.1 ft-lb. Now that both the maximum positive and negative moments have been computed, we may construct the bending moment diagram, Figure 9-3c.

9-3. Continuous Beam with Unequal Spans

The example in the preceding article dealt with a continuous beam having two equal spans. The following example applies the same form of the three-moment equation to the case of unequal spans.

Example. A continuous beam with no restraint at the end supports has spans of 14 ft [4.2 m] and 10 ft [3 m]. A uniformly distributed load of 1000 lb/ft [15 kN/m] extends over its entire length as shown in Figure 9-4a. Construct the shear and bending moment diagrams.

Solution: (1) By data we know that $L_1 = 14$ ft, $L_2 = 10$ ft, w_1 and w_2 each equal 1000 lb per lin ft, and $M_1 = M_3 = 0$. Therefore the first step is to determine M_2 by substituting the known values in the three-

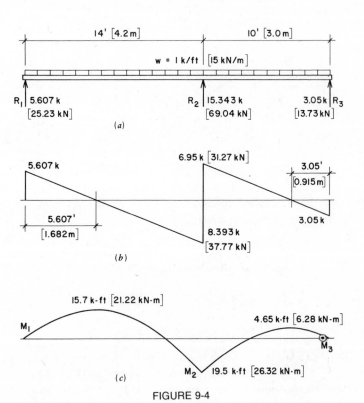

FIGURE 9-4

moment equation

$$M_1L_1 + 2M_2(L_1 + L_2) + M_3L_2 = -\frac{w_1L_1^3}{4} - \frac{w_2L_2^3}{4}$$

Thus

$$0 + 2M_2(14 + 10) + 0 = -\frac{1000 \times 14^3}{4} - \frac{1000 \times 10^3}{4}$$

$$48M_2 = -936,000$$

$$M_2 = -19,500 \text{ ft-lb [26.325 kN-m]}$$

(2) Now let us write an expression for the bending moment at $x = 14$ (at R_2), considering the forces to the left. This expression can be equated to $-19,500$ ft-lb since we have just established its magnitude. Thus

$$14R_1 - (14 \times 1000 \times 7) = -19,500 \quad \text{and}$$

$$R_1 = 5607 \text{ lb [25.232 kN]}$$

Similarly, writing an equation for the bending moment at R_2, considering the forces to the *right*,

$$10R_3 - (10 \times 1000 \times 5) = -19,500 \quad \text{and}$$

$$R_3 = 3050 \text{ lb [13.725 kN]}$$

(3) Computing R_2,

$$R_1 + R_2 + R_3 = (14 \times 1000) + (10 \times 1000)$$

$$5607 + 3050 + R_2 = 24,000 \text{ lb}$$

$$R_2 = 15,343 \text{ lb [69.04 kN]}$$

(4) Now that the values of R_1, R_2, and R_3 have been determined, we can draw the shear diagram, Figure 9-4b.

Let $x =$ the distance from R_1 to the point at which the shear is zero. Then

$$5607 - (1000 \times x) = 0 \quad \text{and} \quad x = 5.607 \text{ ft [1.682 m]}$$

Similarly, we find that the shear is zero in the right-hand span at 3.05 ft [0.915 m] from R_3.

(5) The maximum positive bending moment between R_1 and R_2 is

$$M_{(x = 5.607)} = (5607 \times 5.607) - \left(5.607 \times 1000 \times \frac{5.607}{2}\right)$$

$$= 15,700 \text{ ft-lb } [21.22 \text{ kN-m}]$$

and the maximum positive moment between R_2 and R_3 is

$$(3050 \times 3.05) - \left(3.05 \times 1000 \times \frac{3.05}{2}\right)$$

$$= 4651.25 \text{ ft-lb } [6.279 \text{ kN-m}]$$

The bending moment diagram is shown in Figure 9-4c.

Problem 9-3-A*.   A beam is continuous through two spans and sustains a uniformly distributed load of 2 k/ft [29 kN/m], including its own weight. The span lengths are 12 ft [3.6 m] and 16 ft [4.8 m]. Find the values of the three reactions and construct the complete shear and moment diagrams.

9-4.   Continuous Beam with Concentrated Loads

In the previous examples the loads on the continuous beams were uniformly distributed. Figure 9-5a shows a continuous beam of two spans with a concentrated load on each span. The general shape of the bending moment diagram for this beam is shown in Figure 9-5b. For these conditions, the form of the three-moment equation is

$$M_1 L_1 + 2M_2(L_1 + L_2) + M_3 L_2 = -P_1 L_1^2 [n_1(1 - n_1)(1 + n_1)]$$

$$- P_2 L_2^2 [n_2(1 - n_2)(2 - n_2)]$$

in which the various terms are as shown in Figure 9-5.

Example.   A two-span continuous beam has equal spans of 20 ft [6 m] each. There is a concentrated load of 4000 lb [18 kN] at the center of each span. The ends of the beam at R_1 and R_2 are simply supported. Compute the magnitude of the reactions and construct the shear and bending moment diagrams. See Figure 9-6a.

Solution: (1) For the given spans and loading, note that $L_1 = L_2 = 20$ ft, P_1 and P_2 = 4000 lb each, and both n_1 and n_2 = 0.5. Substituting

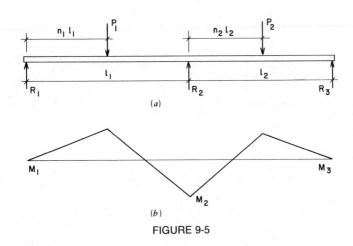

FIGURE 9-5

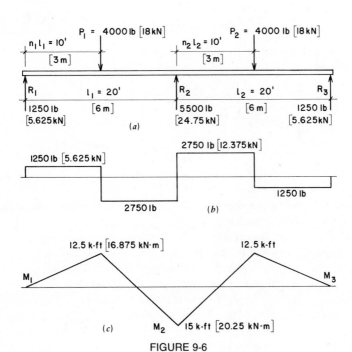

FIGURE 9-6

177

the known values in the form of the three-moment equation stated previously.

$$(M_1 \times 20) + 2M_2(20 + 20) + (M_3 \times 20)$$
$$= -4000 \times 20^2(0.5 \times 0.5 \times 1.5)$$
$$- 4000 \times 20^2(0.5 \times 0.5 \times 1.5)$$

Because there is no restraint at R_1 and R_3, the values of M_1 and M_3 each equal zero. Thus

$$0 + 80M_2 + 0 = -(4000 \times 400 \times 0.375)$$
$$- (4000 \times 400 \times 0.375)$$

or

$$80M_2 = -1,200,000$$

and

$$M_2 = -15,000 \text{ ft-lb } [20.25 \text{ kN-m}]$$

which is the value of the negative bending moment at R_2, the center support.

(2) Next, write an expression for the bending moment at R_2 and equate it to $-15,000$ ft-lb, the value we have just found. Thus

$$M_{(x = 20)} = (R_1 \times 20) - (4000 \times 10) = -15,000$$

$$20R_1 = 25,000 \quad \text{and} \quad R_1 = 1250 \text{ lb } [5.625 \text{ kN}]$$

R_1 and R_3 will have equal magnitudes, hence $R_1 = R_3 = 1250$ lb and $R_2 = (4000 + 4000) - (1250 + 1250) = 5500$ lb [24.75 kN]. Now that all the vertical forces are known, the shear diagram can be constructed. It is shown in Figure 9-6b.

(3) Since the magnitude of R_1 has been established, we can now compute the value of the bending moment under the first concentrated load. Thus $M_{(x = 10)} = 1250 \times 10 = 12,500$ ft-lb [16.875 kN-m]. This is also the magnitude of the bending moment under the second concentrated load. We know that the bending moment has a magnitude of $-15,000$ ft-lb at R_2. Therefore, we have sufficient data to construct the bending moment diagram for this continuous beam; it is shown in Figure 9-6c.

9-5. Continuous Beam with Three Spans

The preceding examples demonstrate that the key operation in design and investigation of continuous beams is the determination of negative bending moment values at the supports. Once these values are established, we can compute the reactions and the positive bending moments. When a continuous beam has more than two spans, the three-moment equation is applied to successive pairs of spans. For the three-span beam shown in Figure 9-7a, the first pair would consist of the span to the left and the middle span; the second, the middle span and the span to the right. This process can be extended to continuous beams with any number of spans.

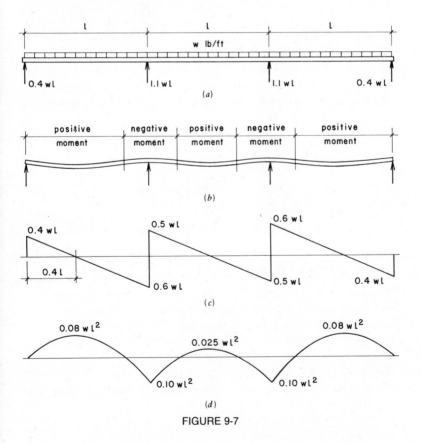

FIGURE 9-7

Figures 9-7c and d represent the shear and bending moment diagrams for the continuous beam shown in Figure 9-7a. The uniformly distributed load is w lb/lin ft and extends the full length of the three equal spans. Successive application of the three-moment equation yields the values recorded in the figure. The values for reactions and shear are coefficients of wL; for the moments the values are coefficients of wL^2. Similar coefficients for continuous beams of various spans and types of loading may be found in engineering reference books (see Section 9-1). Such tabulations enable one to determine quickly the bending moments at critical sections.

Example.   A continuous beam has three spans of 20 ft [6 m] each and a uniformly distributed load of 800 lb/ft [12 kN/m] extending over the entire length of the beam. The ends of the beam are simply supported. Compute the maximum bending moment and the maximum shear.

Solution: (1) Referring to Fig. 9-7d, we note that the maximum positive bending moment $(0.08\,wL^2)$ occurs near the middle of each end span, and the maximum negative moment $(0.10\,wL^2)$ occurs over each of the interior supports. Using the larger value, the maximum bending moment on the beam is

$$M = -0.10\,wL^2 = -(0.10 \times 800 \times 20 \times 20)$$

$$= -32{,}000 \text{ ft-lb [43.2 kN-m]}$$

(2) Figure 9-7c shows that the maximum shear occurs at the face of the first interior support (working from either left or right) and is

$$V = 0.6\,wL = 0.6 \times 800 \times 20 = 9600 \text{ lb [43.2 kN]}$$

9-6.   Restrained Beams

A simple beam was defined in Section 7-1 as a beam that rests on a support at each end, there being no restraint against bending at the supports; the ends are *simply supported*. The shape a simple beam tends to assume under load is shown in Figure 9-8a. Figure 9-8b shows a beam built into a wall at the left end and simply supported at the right end. The left end is *restrained* or *fixed*; the tangent to the elastic curve is horizontal at the face of the wall, at which point there is a negative bending moment. For such a beam, symmetrically loaded, the reactions are not equal in magnitude as they are in simple beams. Figure 9-8c illustrates a beam fixed at both ends. In this beam negative

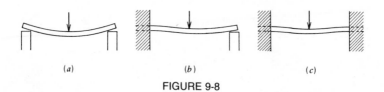

FIGURE 9-8

bending moments occur at the supports, and positive moment occurs in a portion of the beam in the central part of the span. A negative bending moment always occurs at a fixed support.

The reactions for beams with fixed ends cannot be computed by the principle of moments, as in simple beams, and the necessary mathematics involved is beyond the scope of this book. Figure 9-9a, and 9-

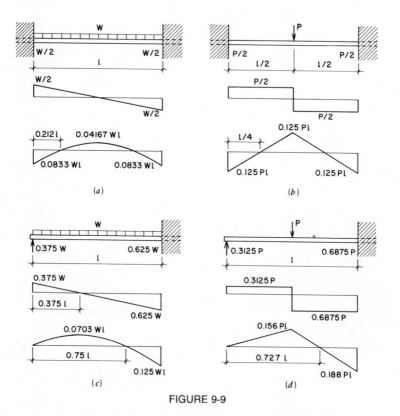

FIGURE 9-9

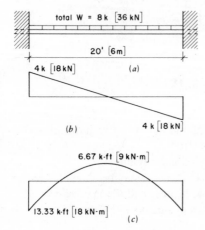

FIGURE 9-10

9b show beams fixed at both ends, and Figure 9-9c and 9-9d indicate the condition where one end is fixed and the other is simply supported. For each of these two cases there are two types of loading—uniformly distributed and concentrated at the center of the span. For each beam, the magnitude of the maximum shear and bending moment is given in the shear and moment diagrams.

Example 1.   Figure 9-10a represents a beam with both ends fixed and having a span of 20 ft [6 m]. It supports a total load of 8000 lb [36 kN] uniformly distributed over its entire length. Draw the shear and bending moment diagrams.

Solution: (1) Referring to Figure 9-9a, we find that the shear at each end, and consequently each reaction, is $W/2$ or $8000 \div 2 = 4000$ lb [18 kN]. Since the load is uniformly distributed, the shear diagram takes the shape shown in Figure 9-10b.

(2) Consulting Figure 9-9 for moment coefficients, we find that the negative bending moment at each support is

$$M = -\frac{WL}{12} = -\frac{8000 \times 20}{12} = -13,330 \text{ ft-lb [18 kN-m]}$$

and the positive moment at the center of the span is

$$M = \frac{WL}{24} = \frac{8000 \times 20}{24} = 6667 \text{ ft-lb [9 kN-m]}$$

The bending moment diagram is shown in Figure 9-10c.

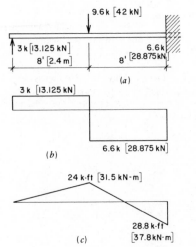

$3 k$ [13.125 kN]

6.6 k [28.875 kN]

24 k·ft [31.5 kN·m]

28.8 k·ft [37.8kN·m]

(c) FIGURE 9-11

Example 2. A beam having a span of 16 ft [4.8 m] is fixed at one end and simply supported at the other (Figure 9-11*a*). There is a load of 9600 lb [42 kN] placed at the center of the span. Draw the shear and bending moment diagrams.

Solution: (1) Shear and moment coefficients for this beam and loading are given in Figure 9-9*d*. The shear at the simply supported end is

$$V = \tfrac{5}{16} \times P = \tfrac{5}{16} \times 9600 = 3000 \text{ lb } [13.125 \text{ kN}]$$

and the shear at the fixed end is

$$V = -(\tfrac{11}{16} \times P) = -(\tfrac{11}{16} \times 9600) = -6600 \text{ lb } [28.875 \text{ kN}]$$

These two values enable us to draw the shear diagram shown in Figure 9-11*b*.

(2) The maximum positive bending moment is

$$M = \tfrac{5}{32} \times Pl = \tfrac{5}{32} \times 9600 \times 16 = 24{,}000 \text{ ft-lb } [31.5 \text{ kN-m}]$$

and the maximum negative moment at the fixed end is

$$M = -(\tfrac{6}{32} \times Pl) = -(\tfrac{6}{32} \times 9600 \times 16)$$

$$= -28{,}800 \text{ ft-lb } [37.8 \text{ kN-m}]$$

The bending moment diagram is shown in Figure 9-11*c*.

Problem 9-6-A.   A beam having a span of 22 ft [6.6 m] is fixed at both ends and has a concentrated load of 16 kips [71 kN] at the center of the span. Compute the maximum shear and bending moment, locate the inflection points, and draw the shear and moment diagrams.

Problem 9-6-B *.   A beam fixed at one end and simply supported at the other has a span of 20 ft [6 m]. There is a total load of 8 kips [36 kN] uniformly distributed over the full span. Compute the maximum shear and bending moment, locate the inflection point, and draw the shear and moment diagrams.

10

Deflection of Beams

‖‖

10-1. General

When a beam bends under load, the deformation or change in shape is called *deflection*. The vertical distance moved by a point on the neutral surface is the deflection of the beam at that point. The trace of the neutral surface on a vertical longitudinal plane is called the *elastic curve* of the beam.

In the design of most beams it is important that the deflection be given consideration. A floor beam may be strong enough to support the load it is required to carry, but it may not be *stiff* enough; that is, the deflection may be so great that a plaster ceiling below might develop cracks or the general lack of stiffness might result in an excessively springy floor. For average conditions the allowable deflection of beams in building construction may be limited to $\frac{1}{360}$ of the span. In the past this limit applied to the total load on a beam but many building codes now permit this allowable deflection under the action of live load only. (*Live load* is the probable load due to occupancy of a building, as distinguished from *dead load* which represents the weight of the construction.) Thus it becomes necessary for the designer, having determined the size of a beam to withstand shear and bending stresses, to compute the actual deflection and check this value against the allowable deflection.

FIGURE 10-1

10-2.   Deflection Formulas

The general form of equations for the deflection of beams is

$$D = K \times \frac{Pl^3}{EI}$$

in which P represents the magnitude of the load, l is the span length, and K is a coefficient related to the distribution of the load (uniform, concentrated, etc.). The modulus of elasticity E is a measure of the stiffness of the material comprising the beam, and the moment of inertia I accounts for the size and shape of the cross section. Examination of this expression shows that deflection increases with an increase in load or span length and decreases with an increase in E or I. Deflection formulas may be derived by methods involving the calculus but we consider only the *moment-area method* which does not require use of the calculus. Three frequently used formulas are derived from the moment-area principle.

The curved line in Figure 10-1 represents a portion of the elastic curve of a beam that was originally straight and horizontal. M and N are two points on the curve, and D represents the vertical displacement of point M from the tangent to the elastic curve at N. According to the moment-area principle, *D is numerically equal to the statical moment about M of the area of the bending moment diagram between M and N, divided by EI.* Application of this principle is demonstrated in the following sections.

10-3.   Cantilever Beam with Load at Free End

Figure 10-2*a* shows a cantilever beam of length l with a concentrated load P at the free end. The elastic curve for this beam is shown in Figure 10-2*b* and the bending moment diagram in Figure 10-2*c*. Note that the tangent to the elastic curve is horizontal at the support and that the maximum deflection is D at the free end. Since the maximum bend-

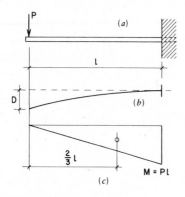

FIGURE 10-2

ing moment occurs at the face of the wall and has a magnitude of Pl, the *area* of the moment diagram between the force P and the face of the wall is $(Pl \times l)/2$ or $Pl^2/2$. Observing that the centroid of the triangular moment diagram is located $2l/3$ distance from the free end of the beam, the statical moment about the free end is $(Pl^2/2) \times (2l/3)$. Therefore, applying the moment-area principle,

$$D = \frac{Pl^2}{2} \times \frac{2l}{3} \times \frac{1}{EI} = \frac{Pl^3}{3EI}$$

which is the formula for maximum deflection of a cantilever beam under this type of loading.

10-4. Simple Beam with Concentrated Load at Midspan

Figure 10-3*a* shows a simple beam of length l with a concentrated load at the center of the span. Figure 10-3*b* and 10-3*c* shows the elastic curve and the bending moment diagram respectively. By observation it is apparent that the maximum deflection occurs at midspan, and consequently the tangent to the elastic curve at that point is a horizontal line. This is true of all symmetrically loaded simple beams. The vertical distance D from this tangent to the intersection of the elastic curve with the support is equal to the deflection at midspan. If we consider this point of intersection (although it does not move) as point M in the statement of the moment-area principle given in Section 10-2, we may apply the principle to this type of beam and loading.

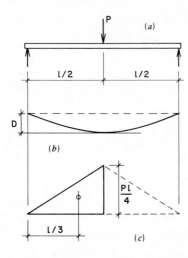

FIGURE 10-3

The area of the moment diagram between the center of the span and the left reaction (outlined area) is $(Pl/4) \times (l/2) \times \frac{1}{2} = Pl^2/16$, and the distance of the centroid of the area from the left reaction is $\frac{2}{3} \times (l/2) = l/3$. Hence the statical moment of this area about the left support is $(Pl^2/16) \times (l/3)$. Applying the moment-area principle,

$$D = \frac{Pl^2}{16} \times \frac{l}{3} \times \frac{1}{EI} = \frac{Pl^3}{48EI}$$

which is the formula for maximum deflection of a simple beam under this type of concentrated loading.

10-5. Simple Beam with Uniformly Distributed Load

One of the most common conditions is a single beam with a uniformly distributed load over its entire length. Such a beam is shown in Figure 10-4a. The elastic curve and bending moment diagram are shown in Figure 10-4b and c, respectively. The curve of the bending moment diagram is a parabola, and the maximum deflection occurs at the center of the span.

The area of the bending moment diagram between midspan and the left reaction is that of the half parabola indicated by outlining in Figure 10-4c. It is equal to two-thirds the product of the height times the base,

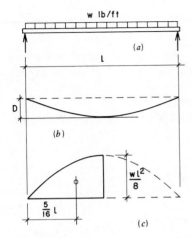

FIGURE 10-4

or $\frac{2}{3} \times (wl^2/8) \times (l/2)$. The distance of the centroid of the half parabola from the left support is five-eighths of the base, or $\frac{5}{8} \times (l/2) = 5l/16$. Applying the moment-area principle to this symmetrically loaded simple beam,

$$D = \frac{2}{3} \times \frac{wl^2}{8} \times \frac{l}{2} \times \frac{5l}{16} \times \frac{1}{EI}$$

$$= \frac{5wl^4}{384 EI} \quad \text{or} \quad \frac{5Wl^3}{384 EI}$$

which express the maximum deflection of a simple beam with a uniformly distributed load over the full span.

10-6. Examples of Deflection Computations

Simple and cantilever beams with cases of typical loadings are shown in Figure 8-10. For each of the six cases the value of the maximum deflection is given in the right-hand column in units of inches. It is important to remember, in using these values, that l, *the length of the span*, *is in inches*.

In designing beams the designer establishes a beam whose size is ample to resist bending stresses; that is, he selects a beam that is *strong* enough to support the applied loads. Having done this, he computes the actual deflection of the beam to see whether it exceeds the allow-able deflection.

Example 1.   An S 12 × 31.8 is to be used as a simple beam on a span of 18 ft [5.4 m]. There will be a total uniformly distributed load of 26 kips [115 kN], including the beam weight. Is the deflection excessive if the allowable limit is $\frac{1}{360}$ of the span?

Solution: (1) Referring to Figure 8-10, Case 2, we find that the maximum deflection is given by the expression $D = 5WL^3/384\,EI$. In this example $W = 26$ kips and $L = 18 × 12 = 216$ in. The moment of inertia of the beam is 218 in.⁴ [90.73 × 10⁶ mm⁴] (Table 6-2), and E for steel will be taken as 29,000 ksi [200 GPa]. Then the computed deflection is

$$D = \frac{5}{384} \times \frac{WL^3}{EI} = \frac{5}{384} \times \frac{26 \times (216)^3}{29,000 \times 218} = 0.54 \text{ in. [13 mm]}$$

(2) The allowable deflection is 18 × 12/360 = 0.6 in. [15 mm]. Since the computed deflection is less than the allowable, the section is acceptable with respect to deflection.

Example 2.   An 8 × 10 beam of Douglas Fir, Select Structural grade, is to carry a concentrated load of 4 kips [18 kN] at the center of a 16-ft [4.8 m] span. Investigate this beam for deflection if D is limited to $L/360$. Ignore the effect of the beam weight.

Solution: (1) Referring to Case 1 of Figure 8-10, we find that the deflection formula is $D = PL^3/48\,EI$. The moment of inertia of a nominal 8 × 10 beam is 536 in.⁴ [223 × 10⁶ mm⁴] from Table 6-7, and the modulus of elasticity for this species and grade of timber is 1600 ksi [11 GPa] from Table 5-3. Using these values, the deflection computation is

$$D = \frac{1}{48} \times \frac{PL^3}{EI} = \frac{1}{48} \times \frac{4 \times (192)^3}{1600 \times 536} = 0.69 \text{ in. [16.9 mm]}$$

(2) The allowable deflection is 16 × 12/360 = 0.53 in. [13.5 mm]. Since the computed deflection exceeds the limit, the 8 × 10 beam is *not* acceptable.

Example 3.   A W 16 × 36 is used for a simple beam on a span of 21 ft [6.3 m]. There are two concentrated loads of 4 kips each [18 kN] located at the third points of the span, and a uniformly distributed load of 800 lb/ft [12 kN/m] extending over the entire span and including the beam weight. Investigate for the deflection under this combined loading if the limit is $L/360$.

Solution: (1) We observe that the beam is symmetrically loaded and that the maximum deflection for both the concentrated and the uniformly distributed loadings occurs at the center of the span. The total deflection at midspan is the sum of the deflections produced separately by each type of loading. The formulas involved are those given for Case 2 and Case 3 in Figure 8-10.

(2) From Table 6-1 we find that I for a W 16 × 36 is 448 in.[4] [186.5 × 10^6 mm^4]. The span in inches is 21 × 12 = 252 in., and E for steel is 29,000 ksi [200 GPa]. Using these values, the deflection due to the concentrated loads is

$$D = \frac{23}{648} \times \frac{PL^3}{EI} = \frac{23}{648} \times \frac{4 \times (252)^3}{29{,}000 \times 448} = 0.175 \text{ in. } [4.28 \text{ mm}]$$

and the deflection caused by the uniform load is

$$D = \frac{5}{384} \times \frac{WL^3}{EI} = \frac{5}{384} \times \frac{21 \times 0.8 \times (252)^3}{29{,}000 \times 448}$$

$$\doteq 0.27 \text{ in. } [6.60 \text{ mm}]$$

making the total deflection $0.175 + 0.27 = 0.445$ in. [10.88 mm].

(3) Since the allowable limit is 21 × 12/360 = 0.7 in. [17.8 mm], the section is acceptable with respect to deflection.

Problem 10-6-A.   A 6 × 12 beam of Southern Pine, No. 1 SR grade, is used on a simple span of 18 ft [5.4 m]. If the total uniformly distributed load is 5 kips [22 kN], and the deflection is limited to $L/360$, is the actual (computed) deflection excessive?

Problem 10-6-B*.   A cantilever beam 8 ft [2.4 m] long has a concentrated load of 6 kips [27 kN] at the free end. If the beam is a W 10 × 33, compute the maximum deflection due to the concentrated load only.

Problem 10-6-C.   An S 12 × 31.8 is used as a cantilever beam having a length of 8 ft [2.4 m]. The beam supports a concentrated load of 6 kips [27 kN] at its free end and also a uniformly distributed load of 3.2 kips (total) [14 kN] that extends over its full length. Compute the maximum deflection.

Problem 10-6-D*.   A 6 × 12 beam of Douglas Fir, No. 1 grade, is used on a simple span of 12 ft [3.6 m]. If the total uniformly distributed load on the beam is 8.5 kips [38 kN], and the deflection is limited to $L/360$, is the actual (computed) deflection excessive?

11

Bending Stresses: Design of Beams

IIIIIIIIIIIIIIIIIIIIIIIIIIIIIIIIIIIIIIIIIII

11-1. Introduction

A beam was defined in Section 7-1 as a structural member that supports transverse loads. The loads tend to produce *bending* rather than shortening or lengthening of the member. In Chapter 7 we found that a beam must resist external shearing forces, and that these forces are held in equilibrium by resisting shearing stresses exerted by the fibers of the beam. We now investigate the *bending stresses* set up in the fibers of a beam that resist the bending moment (Section 8-1) caused by the external forces on the beam.

Figure 11-1a represents the side elevation of a simple beam before it is affected by any loads; the upper and lower surfaces are plane parallel surfaces. The horizontal dot-and-dash line is the neutral surface, and *AB* and *CD* are two parallel lines drawn on the side of the beam. Figure 11-1b represents the same side elevation of the beam when it is deformed (bent) by applied loads. Points *A* and *C* have moved closer together, and points *B* and *D* are farther apart. The fibers of the beam above the neutral surface are in compression and those below are in tension. The maximum compressive stress is in a fiber at the upper surface of the beam. The stresses below the upper surface decrease in magnitude and are zero at the neutral surface. Similarly, the stresses

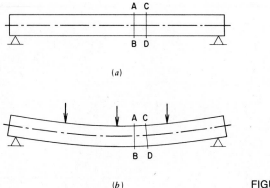

FIGURE 11-1

below the neutral surface are tensile, the maximum stress being at the bottom surface of the beam. If the stress in the outermost fibers of the bent beam does not exceed the elastic limit of the material, the lines *AB* and *CD* remain straight lines. The deformation (lengthening and shortening) of the fibers, and also the magnitude of the stresses in the fibers, is directly proportional to their distances from the neutral surface.

11-2. Resisting Moment

We learned in the preceding chapter that bending moment is a measure of the tendency of the external forces on a beam to deform it by bending. We now consider the action within the beam that resists bending and is called the *resisting moment*.

Figure 11-2*a* shows a simple beam, rectangular in cross section, supporting a single concentrated load *P*. Figure 11-2*b* is an enlarged sketch of the left-hand portion of the beam between the reaction and section *X-X*. From the preceding discussions we know that the reaction R_1 tends to cause a clockwise rotation about point *A* in the section under consideration; this we have defined as the bending moment at the section. In this type of beam the fibers in the upper part are in compression, and those in the lower part are in tension. There is a horizontal plane separating the compressive and tensile stresses; it is called the *neutral surface*, and at this plane there are neither compres-

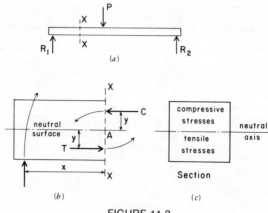

FIGURE 11-2

sive nor tensile stresses with respect to bending. The line in which the neutral surface intersects the beam cross section (Figure 11-2c) is called the *neutral axis*, NA.

Call C the sum of all the compressive stresses acting on the upper part of the cross section, and call T the sum of all the tensile stresses acting on the lower part. It is the sum of the moments of these stresses at the section that holds the beam in equilibrium; this is called the *resisting moment* and is equal to the bending moment in magnitude. The bending moment about point A is $R_1 \times x$, and the resisting moment about the same point is $(C \times y) + (T \times y)$. The bending moment tends to cause a clockwise rotation, and the resisting moment tends to cause a counterclockwise rotation. If the beam is in equilibrium, these moments are equal, or

$$R_1 \times x = (C \times y) + (T \times y)$$

That is, the bending moment equals the resisting moment. This is the theory of flexure (bending) in beams. For any type of beam, we can compute the bending moment; and if we wish to design a beam to withstand this tendency to bend, we must select a member with a cross section of such shape, area, and material that it is capable of developing a resisting moment equal to the bending moment.

11-3.   The Flexure Formula

The flexure formula $M = fS$ is an expression for resisting moment that involves the size and shape of the beam cross section (represented by S in the formula) and the material of which the beam is made (represented by f). It is used in the design of all homogeneous beams, that is, beams made of one material only, such as steel or wood. You will never need to derive this formula, but you will use it many times. The following brief derivation is presented to show the principles on which the formula is based.

Figure 11-3 represents a partial side elevation and the cross section of a homogeneous beam subjected to bending stresses. The cross section shown is unsymmetrical about the neutral axis, but this discussion applies to a cross section of any shape. In Figure 11-3a let c be the distance of the fiber farthest from the neutral axis, and let f be the unit stress on the fiber at distance c. If f, the extreme fiber stress, does not exceed the elastic limit of the material, the stresses in the other fibers are directly proportional to their distances from the neutral axis. That is to say, if one fiber is twice the distance from the neutral axis than another fiber, the fiber at the greater distance will have twice the stress. The stresses are indicated in the figure by the small lines with arrows, which represent the compressive and tensile stresses acting toward and away from the section, respectively. If c is in inches, the unit stress on a fiber at 1 in. distance is f/c. Now imagine an infinitely small area a at z distance from the neutral axis. The unit stress on this fiber is $(f/c) \times z$, and because this small area contains a square inches, the

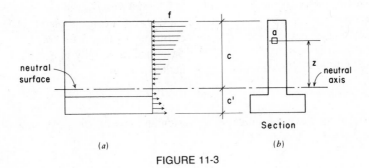

FIGURE 11-3

total stress on fiber a is $(f/c) \times z \times a$. The *moment* of the stress on fiber a at z distance is

$$\frac{f}{c} \times z \times a \times z \quad \text{or} \quad \frac{f}{c} \times a \times z^2$$

We know, however, that there is an extremely large number of these minute areas, and if we use the symbol Σ to represent the sum of this very large number, we can write

$$\Sigma \frac{f}{c} \times a \times z^2$$

which means the sum of the moments of all the stresses in the cross section with respect to the neutral axis. This we know is the *resisting moment*, and it is equal to the bending moment.

Therefore

$$M = \frac{f}{c} \Sigma a \times z^2$$

The quantity $\Sigma a \times z^2$ may be read "the sum of the products of all the elementary areas times the square of their distances from the neutral axis." We call this the *moment of inertia* and represent it by the letter I. Therefore, substituting in the above, we have

$$M = \frac{f}{c} \times I \quad \text{or} \quad M = \frac{fI}{c}$$

This is known as the *flexure formula* or *beam formula*, and by its use we may design any beam that is composed of a single material. The expression may be simplified further by substituting S for I/c, called the *section modulus*, a term that is described more fully in Section 6-5. Making this substitution, the formula becomes

$$M = fS$$

Use of the flexure formula in the design and investigation of beams is discussed in Section 11-5.

11-4.   Moment of Inertia of a Rectangle

It is stated in Section 6-3 that the moment of inertia I of a rectangular cross section with respect to an axis through its centroid and parallel to its base is $bd^3/12$. Also, as shown in Section 6-5, the section modulus S of a rectangular cross section is $bd^2/6$. By use of the flexure formula $M = fS$ we can verify these values.

Figure 11-4a shows a rectangle of width b and depth d, which represents the cross section of a rectangular beam. The neutral axis of the section lies at a distance $d/2$ from both the upper and lower surfaces of the beam, making $c = d/2$. The area in compression (hatched) lies above the neutral axis and is equal to $b \times d/2$. Since the unit stress on the most remote fiber is f and is zero at the neutral axis, the *average* unit stress is $f/2$, and the total compressive stress is $(b \times d/2) \times f/2$.

The resultant of the compressive stresses acts at a distance $\frac{2}{3} \times d/2$ $= d/3$ above the neutral axis (Fig. 11-4b). Therefore the moment of all the compressive stresses with respect to the neutral axis is

$$\left(b \times \frac{d}{2}\right) \times \frac{f}{2} \times \frac{d}{3} = f \times \frac{bd^2}{12}$$

Since there are equal tensile stresses below the neutral axis tending to cause rotation about the axis in the same direction, the sum of the moments of *all* the stresses is $2 \times f \times bd^2/12 = f \times bd^2/6$. This, we know, is the resisting moment. Equating this expression to the bending

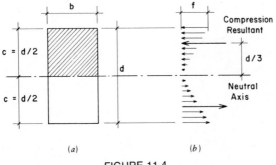

(a)   (b)

FIGURE 11-4

moment,

$$M = f \times \frac{bd^2}{6} \quad \text{or} \quad \frac{M}{f} = \frac{bd^2}{6}$$

The flexure formula may be written $M/f = S$; therefore $S = bd^2/6$, the *section modulus* of the rectangle.

The section modulus may also be expressed as $S = I/c$ or $I/c = bd^2/6$. Since we know that $c = d/2$, $I = (bd^2/6) \times d/2 = bd^3/12$, which is the value of the *moment of inertia* of the rectangular cross section.

11-5. Investigation of Beams

Another use of the flexure formula is for the investigation of beams. To investigate a beam, we determine whether it is strong enough to support a certain loading. For such problems we are given as data a span length, magnitude and type of loading, and the size of the beam. Although such problems may be solved by different methods, a simple and direct way is to compute the *actual* extreme fiber stress in the member and to compare it with the stress we know to be the allowable. This means that we compute the value of f in the flexure formula.

Another method of investigation is to compute the required section modulus for the given load and span so that it may be compared with that of the given section.

Example 1. A W 10 × 26 is proposed to carry a total uniformly distributed load of 30 kips [133 kN], including an allowance for the beam's weight, on a span of 13 ft [4 m]. (See Figure 11-5.) If the allowable bending stress is 24 ksi [165 MPa], determine whether the beam is safe (1) by comparing the maximum resisting moment of the section with the maximum bending moment developed by the loading

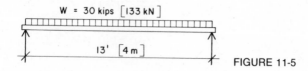

FIGURE 11-5

and (2) by comparing the allowable bending stress with that actually produced by the loading.

Solution (a): By referring to Table 6-1, we find that the section modulus for the W 10 × 26 is 27.9 in.³ [457 × 10³ mm³]. Then

$$M = F_bS = 24 \times 27.9 = 670 \text{ kip-in.}$$

or

$$M = \frac{670}{12} = 55.8 \text{ kip-ft}$$

$$\left[M = F_bS = \frac{165 \times 457 \times 10^3}{10^6} = 75.4 \text{ kN-m} \right]$$

From Section 8-11 and Figure 8-10 (Case 2) we know that the maximum bending moment for the proposed loading occurs at midspan and may be found with the formula $M = WL/8$. Then

$$M = \frac{WL}{8} = \frac{30 \times 13}{8} = 48.8 \text{ kip-ft}$$

$$\left[M = \frac{133 \times 4}{8} = 66.5 \text{ kN-m} \right]$$

The beam is safe as long as the bending moment developed by the loading is less than the permissible resisting moment.

Solution (b): The maximum bending stress will occur at the top and bottom surfaces (of this symmetrical section) at the location of the largest bending moment. Then

$$f_b = \frac{M}{S} = \frac{48.8 \times 12}{27.9} = 20.99 \text{ ksi}$$

$$\left[f_b = \frac{66.5 \times 10^6}{457 \times 10^3} = 146 \text{ MPa} \right]$$

This equation verifies that the beam is safe because the actual extreme fiber stress (20.99 ksi [146 MPa]) is less than the allowable (24 ksi [165 MPa]). Note that the bending moment was multiplied by 12 to convert it to kip-inches.

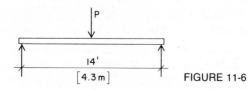

FIGURE 11-6

Example 2. An S 12 × 31.8 has a span of 14 ft [4.3 m]. If the allowable bending stress is 22,000 psi [152 MPa], find the maximum concentrated load it will support at midspan (Figure 11-6).

Solution: By referring to Table 6-3 we find that the section modulus for the beam is 36.4 in.^{3} [597 × 10^3 mm^3]. The maximum resisting moment of the beam is

$$M = F_b S = 22,000 \times 36.4 = 800,800 \text{ in.-lb}$$

or

$$M = \frac{800,800}{12} = 66,733 \text{ ft-lb}$$

$$\left[M = \frac{152 \times 597 \times 10^3}{10^6} = 90.7 \text{ kN-m} \right]$$

From Section 8-11 and Figure 8-10 (Case 1) we find that the maximum bending moment for this loading occurs at midspan and is given by the formula $M = PL/4$. Before this equation is solved for P, however, the bending moment due to the beam weight must be deducted from the maximum resisting moment. The moment due to the beam weight of 31.8 lb/ft [0.464 kN/m] also occurs at midspan and is found from the expression $M = wL^2/8$ (Section 8-6). Then

$$M = \frac{wL^2}{8} = \frac{31.8 \times (14)^2}{8} = 779 \text{ ft-lb}$$

$$\left[M = \frac{0.464 \times (4.3)^2}{8} = 1.07 \text{ kN-m} \right]$$

and the resisting moment available to support the proposed loading is

$$M = 66,733 - 779 = 65,954 \text{ ft-lb}$$

$$[M = 90.7 - 1.1 = 89.6 \text{ kN-m}]$$

Therefore the maximum safe load at the center of the span is

$$P = \frac{4M}{L} = \frac{4 \times 65,954}{14} = 18,844 \text{ lb}$$

$$\left[P = \frac{4 \times 89.6}{4.3} = 83.3 \text{ kN} \right]$$

Problem 11-5-A. *   An S 10 $\times$ 25.4 has a span of 10 ft [3 m] with a uniformly distributed load of 36 kips [160 kN] in addition to its own weight. The allowable bending stress is 24 ksi [165 MPa]. Is the beam safe with respect to bending stresses?

Problem 11-5-B.   A W 16 $\times$ 45 has a loading consisting of 10 kips [45 kN] at each of the quarter points of a 24 ft [7.2 m] span (Figure 8-10, Case 5) and a uniformly distributed load of 5.2 kips [23 kN] including the beam weight. If the allowable bending stress is 24 ksi [165 MPa], is the beam safe with respect to bending stresses?

Problem 11-5-C. *   Two 5 $\times$ 3.5 $\times$ 0.5 in. angles fastened together back-to-back are to be used as a beam on a span of 5 ft [1.5 m]. The allowable bending stress is 22 ksi [152 MPa]. Find the total permissible uniformly distributed load (1) when the long legs are placed vertically back-to-back and (2) when the short legs are so placed.

11-6.   Computation of Safe Loads

The flexure formula can also be used to determine the allowable load a beam of given cress section and span will properly support. Since the flexure formula is $M = fS$, the allowable load is found by expressing the allowable maximum bending moment (the resisting moment) in terms of W or P.

Example 1.   A W 12 $\times$ 40 is used as a simple beam on a span of 14 ft. What is the maximum uniformly distributed load this beam will support?

Solution: (1) From Table 6-1 we find that the section modulus is 51.9 in.^{3} The maximum bending moment for this loading is $M = WL/8$ (Figure 8-10, Case 2).

(2) The bending moment expressed in terms of W is

$$M = \frac{WL}{8} = \frac{W \times 14 \times 12}{8} = 21W \text{ kip-in.}$$

(3) Taking f as 24 ksi, and substituting in the flexure formula, we

determine the allowable uniformly distributed load to be

$$f = 24 = \frac{M}{S} = \frac{21W}{51.9} \quad \text{and} \quad W = \frac{51.9 \times 24}{21} = 59.3 \text{ kips}$$

Example 2.   A simple beam having a span of 20 ft has a concentrated load applied 12 ft from the left reaction. If the section used is a W 16 $\times$ 36, find the greatest concentrated load it will support, neglecting the effect of the beam weight.

Solution: (1) Using the principle of inverse proportion for the beam with a single load, we find the left reaction will be equal to $\frac{8}{20}$ of the load, or $R_1 = 0.4P$, and the right reaction will be $\frac{12}{20}$ of the load, or $R_2 = 0.6P$.

(2) The moment may be found by taking the product of either reaction about the load point. Thus

$$M = R_1 \times 12 = 0.4P \times 12 = 4.8P \text{ k-ft}$$

or

$$4.8P \times 12 = 57.6P \text{ k-in.}$$

and for a check

$$M = R_2 \times 8 = 0.6P \times 8 = 4.8P \text{ k-ft}$$

(3) Referring to Table 6-1, we find the section modulus to be 56.5 in.^{3}. Taking the allowable bending stress as 24 ksi (Table 5-2) and substituting, we determine

$$f = 24 = \frac{M}{S} = \frac{57.6P}{56.5} \quad \text{and} \quad P = \frac{56.5 \times 24}{57.6} = 23.5 \text{ kips}$$

Note: In solving the following problems, neglect the weight of the beam and determine the allowable loads with respect to bending strength only.

Problem 11-6-A.   Compute the maximum allowable uniformly distributed load for a simple beam with a span of 16 ft [4.8 m] if the section used is an S 12 $\times$ 31.8.

Problem 11-6-B.   An 8 $\times$ 12 wood beam, for which the allowable bending stress is 1400 psi [9.65 MPa], has a span of 15 ft [4.5 m] with equal concentrated loads at the third points of the span. Compute the maximum permitted magnitude for the loads.

Problem 11-6-C*.   A W 14 × 30 having a span of 14 ft [4.2 m] supports a uniformly distributed load of 7 kips [31 kN] and also a concentrated load at the center of the span. Compute the magnitude of the maximum allowable concentrated load.

Problem 11-6-D.   What is the maximum concentrated load that may be placed at the free end of a cantilever beam 9 ft [2.7 m] long if the section used is a W 12 × 22?

Problem 11-6-E.   A simple beam has a span of 20 ft [6 m] with a concentrated load applied 4 ft [1.2 m] from one of the supports. If the section used is a W 16 × 36, compute the allowable magnitude of the concentrated load.

11-7.   Design of Beams

The flexure formula is used primarily to determine the size of a beam with respect to *strength in bending*. Experience has shown that beams that fail usually fail by a crushing or tearing of the fibers at the section at which the bending moment is maximum rather than by shear at the supports. The common procedure in designing a beam is to begin by determining its size in accordance with strength in bending, the application of the flexure formula. The beam selected is then investigated for shear and deflection. Short beams with relatively large loads should always be tested with respect to shear. We known how to compute the maximum bending moment, and now that we have the flexure formula $S = M/f$, the design of beams with respect to strength in bending is a relatively simple matter.

It should be noted that the flexure formula is to be used for *homogeneous beams* only; beams composed of only one material such as steel or wood. It is not applicable to reinforced concrete beams.

Example 1.   A simple beam having a span of 22 ft [6.6 m] is to support a uniformly distributed load of 36 kips [160 kN] including the beam weight. If allowable bending stress is 24 ksi [165 MPa], design a steel beam for strength in bending.

Solution: (1) To apply the flexure formula, we must first determine the maximum bending moment. From Figure 8-10, for Case 2, we find that $M = WL/8$. Then

$$M = \frac{WL}{8} = \frac{36 \times 22}{8} = 99 \text{ kip-ft} \quad \text{or} \quad 1188 \text{ kip-in. [132 kN-m]}$$

(2) Applying the flexure formula, we find the required section modulus is to be

$$S = \frac{M}{f} = \frac{1188}{24} = 49.5 \text{ in.}^3 \ [800 \times 10^3 \text{ mm}^3]$$

(3) Referring to Table 6-1 we find that a $W \ 16 \times 36$ has a section modulus of 56.5 in.^{3} and is therefore acceptable. Other sections having a section modulus of at least 49.5 in.^{3} are also acceptable; for example, a $W \ 14 \times 43$ has an S of 62.7 in.^{3} and an $S \ 15 \times 42.9$ (Table 6-2) has an S of 59.6 in.^{3}. If there is no restriction on beam depth, the lightest-weight section is usually the most economical.

Example 2. A simple beam has a span of 16 ft [4.8 m] and supports a load, including its own weight, of 6.5 kips [28.9 kN]. If the wood to be used for the beam is Douglas Fir, Select Structural grade, determine the size of the beam with the least cross-sectional area on the basis of limiting bending stress.

Solution: The maximum bending moment is

$$M = \frac{Wl}{8} = \frac{6.5 \times 16}{8} = 13.0 \text{ kip-ft } [17.34 \text{ kN-m}]$$

Referring to Table 5-3, we find under Douglas Fir, beams and stringers, Select Structural grade, that the limiting bending stress is 1600 psi [11.03 MPa]. Then, substituting in the flexure formula, we can compute the required section modulus as

$$S = \frac{M}{F_b} = \frac{13,000 \times 12}{1600} = 97.5 \text{ in.}^3 \ [1.57 \times 10^6 \text{ mm}^3]$$

From Table 6-7 we find the section with the least area to be a 4 × 14 with $S = 102.411$ in.^{3} [1.68 × 10^6 mm^3]. Actually this section falls into the higher stress category for sections 2–4 in. thick and 5 in. and wider, which merely indicates that the allowable stress is 1800 psi. It is still the section with the least area.

Example 3. A simple beam with a span of 14 ft [4.2 m] has a concentrated load of 7.7 kips [34 kN] placed 4 ft [1.2 m] from the left support. With respect to strength in bending, design both a steel and wood beam to support this load (neglecting the beam weight in each case). Assume 24 ksi [165 MPa] for allowable bending stress in the steel beam and 1200 psi [8.25 MPa] in the wood beam.

Solution: (1) Computing the reactions,

$$R \times 14 = 7.7 \times 10 \quad \text{and} \quad R_1 = 5.5 \text{ k [24.28 kN]}$$

$$R_2 \times 1 = 7.7 \times 4 \quad \text{and} \quad R_2 = 2.2 \text{ k [9.72 kN]}$$

(2) The maximum bending moment occurs at the point of the load and may be found as the moment of either reaction about this point. Thus

$$M = R_1 \times 4 = 5.5 \times 4 = 22 \text{ k-ft [29.14 kN-m]}$$

or

$$M = R_2 \times 10 = 2.2 \times 10 = 22 \text{ k-ft}$$

(3) The required section modulus for the steel beam is

$$S = \frac{M}{f} = \frac{22 \times 12}{24} = 11.0 \text{ in.}^3 \, [177 \times 10^3 \text{ mm}^3]$$

Referring to Table 6-1, we find that a $W \, 8 \times 18$ has a section modulus of 15.2 in.^{3} and is therefore acceptable.

(4) The required section modulus for the wood beam is

$$S = \frac{M}{f} = \frac{22 \times 12}{1.200} = 220 \text{ in.}^3 \, [3532 \times 10^3 \text{ mm}^3]$$

Table 6-7 shows that an 8×14 timber with a section modulus of 227 in.^{3} is acceptable

Note: In the following problems neglect the weight of the beam and determine the proper sections with respect to bending strength only. Use a value of 24 ksi [165 MPa] for the allowable stress for steel.

Problem 11-7-A.   A simple beam has a span of 17 ft [5.1 m] and supports a uniformly distributed load of 23 kips [102 kN]. Determine the size of a wide flange steel beam to carry this load.

Problem 11-7-B.   Two loads of 11 kips [49 kN] each occur at the third points of the span of a simple beam whose length is 18 ft [5.4 m]. What is the lightest-weight standard I-beam (Table 6-2) that is acceptable?

Problem 11-7-C*. A simple beam with a 20-ft [6-m] span has a concentrated load of 20 kips [89 kN] at the center of the span and, in addition, a uniformly distributed load of 200 lb/ft [2.9 kN/m] over its entire length. Determine the lightest-weight wide flange section that is acceptable.

Problem 11-7-D. A wood beam of Douglas Fir, Select Structural grade, is to be used on a span of 15 ft [4.5 m]. If there is a concentrated load of 6.3 kips [28 kN] located 5 ft [1.5 m] from one end, what section is acceptable?

Problem 11-7-E*. A simple beam of Douglas Fir, Select Structural grade, has a span of 18 ft [5.4 m] with two concentrated loads of 3 kips [13.34 kN] each placed at the third points of the span. Determine the size of the beam with the least cross-sectional area.

Problem 11-7-F. A Southern Pine beam of No. 1 SR grade has a span of 15 ft [4.5 m] with a single concentrated load of 6 kips [26.7 kN] placed 5 ft [1.5 m] from one support. Select the beam with the least cross-sectional area.

12

Built-Up Beams of Two Materials

||||||||||||||||||||||||||||||||||||||||||||||||||

12-1. Beams of Two Materials

The discussion of bending stresses presented in Chapter 11 pertains to beams consisting of a single material only; that is, to *homogeneous* beams. Reinforced concrete construction utilizes beams of two materials, steel and concrete, acting together (Chapter 17). Before the advent of glued laminated wood beams, it was common practice to increase the strength of timber beams by the addition of steel plates. Two means of achieving such a built-up beam section are shown in Figure 12-1. This composite steel and wood member is known as a *flitched beam*.

12-2. Flitched Beams

The components of a flitched beam are securely held together with through bolts so that the elements act as a single unit. The computations for determining the strength of such a beam illustrate the phenomenon of two different materials in a beam acting as a unit. The computations are based on the premise that the two materials deform equally. Let

(a)   (b)   FIGURE 12-1

Δ_1 and Δ_2 = the deformations per unit length of the outermost fibers of the two materials, respectively

f_1 and f_2 = the unit bending stresses in the outermost fibers of the two materials, respectively

E_1 and E_2 = the moduli of elasticity of the two materials, respectively

Since, by definition, the modulus of elasticity of a material is equal to the unit stress divided by the unit deformation, then

$$E_1 = \frac{f_1}{\Delta_1} \quad \text{and} \quad E_2 = \frac{f_2}{\Delta_2}$$

and, transposing

$$\Delta_1 = \frac{f_1}{E_1} \quad \text{and} \quad \Delta_2 = \frac{f_2}{E_2}$$

Since the two deformations must be equal,

$$\frac{f_1}{E_1} = \frac{f_2}{E_2} \quad \text{and} \quad f_2 = f_1 \times \frac{E_2}{E_1}$$

This simple equation for the relationship between the stresses in the two materials of a composite beam may be used as the basis for investigation or design of a flitched beam, as is demonstrated in the following example.

Example. A flitched beam is formed as shown in Fig. 12-1a, consisting of two 2 × 12 planks of Douglas Fir, No. 1 grade, and a 0.5 × 11.25-in. [13 × 285 mm] plate of A36 steel. Compute the allowable uniformly distributed load this beam will carry on a simple span of 14 ft [4.2 m].

Solution: (1) We first apply the formula just derived to determine which of the two materials limits the beam action. For this we obtain the following data.

For the steel: $E = 29,000,000$ psi [200 GPa], and the maximum allowable bending stress F_b is 22 ksi [150 MPa] (Table 5-2).

For the wood: $E = 1,800,000$ psi [12.4 GPA], and the maximum allowable bending stress for single-member use is 1500 psi [10.3 MPa] (Table 5-3).

For a trial we assume the stress in the steel plate to be the limiting value and use the formula to find the maximum useable stress in the wood. Thus

$$f_w = f_s \times \frac{E_w}{E_s} = 22,000 \times \frac{1,800,000}{29,000,000} = 1366 \text{ psi } [9.3 \text{ MPa}]$$

As this produces a stress lower than that of the table limit for the wood, our assumption is correct. That is, if we permit a stress higher than 1366 psi in the wood, the steel stress will exceed its limit of 22 ksi.

(2) Using the stress limit just determined for the wood, we now find the capacity of the wood members. Calling the load capacity of the wood W_w, we find

$$M = \frac{W_w L}{8} = \frac{W_w \times 14 \times 12}{8} = 21 W_w$$

Then using the S of 31.6 in.^{3} for the 2 × 12 (Table 6-7), we find

$$M = 21 W_w = f_w \times S_w = 1366 \times (2 \times 31.6)$$
$$W_w = 4111 \text{ lb } [18.35 \text{ kN}]$$

(3) For the plate we first must find the section modulus as follows (See Fig. 6-10.):

$$S_s = \frac{bd^2}{6} = \frac{0.5 \times (11.25)^2}{6} = 10.55 \text{ in.}^3 \ [176 \times 10^3 \text{ mm}^3]$$

Then

$$M = 21 W_s = f_s \times S_s = 22,000 \times 10.55$$
$$W_s = 11,052 \text{ lb } [50.29 \text{ kN}]$$

And the total capacity of the combined section is

$$W = W_w + W_s = 4111 + 11,052 = 15,163 \text{ lb } [68.64 \text{ kN}]$$

Although the load-carrying capacity of the wood elements is actually reduced in the flitched beam, the resulting total capacity is substantially greater than that of the wood members alone. This significant increase in strength achieved with small increase in size is a principal reason for popularity of the flitched beam. In addition, there is a significant reduction in deflection in most applications, and—most noteworthy—a reduction in sag over time.

Problem 12-2-A*. A flitched beam consists of a single 10 × 14 of Douglas Fir, Select Structural grade, and two A36 steel plates, each 0.5 × 13.5 in. [13 × 343 mm]. Compute the magnitude of the concentrated load this flitched beam will support at the center of a 16-ft [4.8-m] simple span. Neglect the weight of the beam. Use a value of 22 ksi for the limiting bending stress in the steel.

13

Columns

||||||||||||||||||||||||||||||||||||||||||||

13-1. Introduction

A column is a compression member, the length of which is several times greater than its least lateral dimension. The term *column* is generally applied to relatively heavy vertical members whereas the term *strut* is given to smaller compression members not necessarily in a vertical position. There are two principal types of wood column. The *simple solid column* consists of a single piece of structural lumber square or rectangular in cross section. Solid columns of circular cross section are also considered simple solid columns but are encountered less frequently. A *spaced column* is an assembly of two pieces separated at the ends and at intermediate points along its length by blocking. In steel construction wide flange structural sections (W shapes) are usually used for columns.

13-2. Wood Columns

The wood column that is used most frequently is the *simple solid column*. Solid columns of round cross section are also considered simple solid columns but are used less frequently. Two other types are *built-up columns*, consisting of multiple solid elements fastened together to form a solid mass, and *glued laminated columns*.

In wood construction the slenderness ratio of a freestanding simple

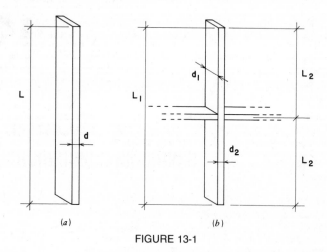

FIGURE 13-1

column is the ratio of the unbraced (laterally unsupported) length to the dimension of its least side, or L/d. (Figure 13-1a.) When members are braced so that the unsupported length with respect to one face is less than that with respect to the other, L is the distance between the points of support that prevent lateral movement in the direction along which the dimension of the section is measured. This is illustrated in Figure 13-1b. If the section is not square or round, it may be necessary to investigate two L/d conditions for such a column to determine which is the limiting one. The slenderness ratio for simple solid columns is limited to $L/d = 50$.

Figure 13-2 illustrates the typical form of the relationship between axial compression capacity and slenderness for a linear compression member (column). The two limiting conditions are those of the very short member and the very long member. The short member (such as a block of wood) fails in crushing, which is limited by the mass of material and the stress limit in compression. The very long member (such as a yardstick) fails in elastic buckling, which is determined by the stiffness of the member; stiffness is determined by a combination of geometric property (shape of the cross section) and material stiffness property (modulus of elasticity). Between these two extremes—which is where most wood compression members fall—the behavior is indeterminate as the transition is made between the two distinctly different modes of behavior.

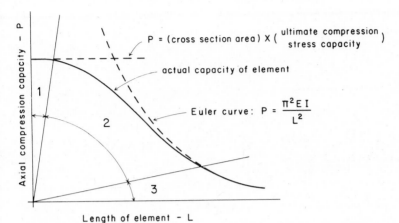

FIGURE 13-2.   Relation of member length to axial compression capacity.

The National Design Specification currently provides for three separate compression stress calculations, corresponding to the three zones of behavior described in Figure 13-2. The plot of these three stress formulas, for a specific example wood, is shown in Fig. 13-3. Typical analysis and design procedures for simple solid wood columns are illustrated in the following examples.

Example 1.   A wood compression member consists of a 3 × 6 of Douglas fir–larch, dense No. 1 grade. Find the allowable axial compression force for unbraced lengths of: (1) 2 ft [0.61 m], (2) 4 ft [1.22 m], (3) 8 ft [2.44 m].
Solution: We find from Table 5-3: F_c = 1450 psi [10.0 MPa] and E = 1,900,000 psi [13.1 GPa]. To establish the zone limits we compute the following:

$$11(d) = 11(2.5) = 27.5 \text{ in.}$$

$$50(d) = 50(2.5) = 125 \text{ in.}$$

and

$$K = 0.671 \sqrt{\frac{E}{F_c}} = 0.671 \sqrt{\frac{1,900,000}{1450}} = 24.29$$

Thus for (1), L = 24 in., which is in Zone 1; $F_c' = F_c = 1450$ psi;

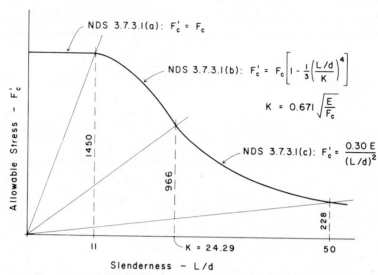

FIGURE 13-3. Allowable axial compression stress as a function of the slenderness ratio L/d. Requirements for Douglas fir-larch, dense No. 1.

allowable $C = F_c' \times$ gross area $= 1450 \times 13.75 = 19{,}938$ lb [88.7 kN].

For (2), $L = 48$ in.; $L/d = 48/2.5 = 19.2$, which is in Zone 2.

$$F_c' = F_c \left\{ 1 - \frac{1}{3} \left(\frac{L/d}{K} \right)^4 \right\}$$

$$= 1450 \left\{ 1 - \frac{1}{3} \left(\frac{19.2}{24.29} \right)^4 \right\}$$

$$= 1262 \text{ psi}$$

Allowable $C = 1262 \times (13.75) = 17{,}353$ lb [77.2 kN].

For (3), $L = 96$ in; $L/d = 96/2.5 = 38.4$, which is in Zone 3.

$$F_c' = \frac{0.3(E)}{(L/d)^2} = \frac{0.3(1{,}900{,}000)}{(38.4)^2} = 387 \text{ psi}$$

Allowable compression $= 387 \times (13.75) = 5321$ lb [23.7 kN].

Example 2. Wood 2 × 4 elements are to be used as vertical compression members to form a wall (ordinary stud wall construction). If the wood is Douglas fir–larch, No. 3 grade, and the wall is 8.5 ft high, what is the column load capacity of a single stud?

Solution: In this case it must be assumed that the surfacing materials used for the wall (plywood, dry wall, plaster, etc.) will provide adequate bracing for the studs on their weak axis (the 1.5-in. [38-mm] direction). If not, the studs cannot be used, since the specified height of the wall is considerably in excess of the limit for L/d for a solid column (50). We therefore assume the direction of potential buckling to be that of the 3.5-in. [89-mm] dimension. Thus

$$\frac{L}{d} = \frac{8.5 \times 12}{3.5} = 29.1$$

In order to determine which column load formula must be used, we must find the value of K for this wood. From Table 5-3 we find $F_c = 675$ psi [4.65 MPa] and $E = 1,500,000$ psi [10.3 GPa]. Then

$$K = 0.671 \sqrt{\frac{E}{F_c}} = 0.671 \sqrt{\frac{1,500,000}{675}} = 31.63$$

We thus establish the condition for the stud as Zone 2 (Figure 13-2), and the allowable compression stress is computed as

$$F'_c = F_c \left\{ 1 - \frac{1}{3} \left(\frac{L/d}{K} \right)^4 \right\}$$

$$= 675 \left\{ 1 - \frac{1}{3} \left(\frac{29.1}{31.63} \right)^4 \right\} = 514 \text{ psi [3.54 MPa]}$$

The allowable load for the stud is

$$P = F'_c \times \text{gross area} = 514 \times 5.25 = 2699 \text{ lb [12.0 kN]}$$

Example 3.   A wood column of Douglas fir–larch, dense No. 1 grade, must carry an axial load of 40 kips [178 kN]. Find the smallest section for unbraced lengths of: (1) 4 ft [1.22 m], (2) 8 ft [2.44 m], (3) 16 ft [4.88 m].

Solution: Since the size of the column is unknown, the values of L/d, F_c, and E cannot be predetermined. Therefore, without design aids (tables, graphs, or computer programs), the process becomes a cut-and-try approach in which a specific value is assumed for d and the resulting values for L/d, F_c, E, and F'_c are determined. A required area is then determined and the sections with the assumed d compared with the requirement. If an acceptable member cannot be found, another try must be made with a different d. Although somewhat clumsy, the process is usually not all that laborious, since a limited number of available sizes are involved.

We first consider the possibility of a Zone 1 stress condition (Figure 13-2), since this calculation is quite simple. If the maximum $L = 11(d)$, then the minimum $d = (4 \times 12)/11 = 4.36$ in. [111 mm]. This requires a nominal thickness of 6 in., which puts the size range into the "posts and timbers" category in Table 5-3, for which the allowable stress F_c is 1200 psi. The required area is thus

$$A = \frac{load}{F_c'} = \frac{40,000}{1200} = 33.3 \text{ in.}^2 \text{ [21,485 mm}^2\text{]}$$

The smallest section is thus a 6 × 8, with an area of 41.25 in^2, since a 6 × 6 with 30.25 in^2 is not sufficient. (See Table 6-7.) If the rectangular-shape column is acceptable, this becomes the smallest member usable. If a square shape is desired, the smallest size would be an 8 × 8.

If the 6-in. nominal thickness is used for the 8-ft column, we determine that

$$\frac{L}{d} = \frac{8 \times 12}{5.5} = 17.45$$

Since this is greater than 11, the allowable stress is in the next zone, for which

$$F_c' = F_c \left\{ 1 - \frac{1}{3} \left(\frac{L/d}{K} \right)^4 \right\}$$

$$= 1200 \left\{ 1 - \frac{1}{3} \left(\frac{17.45}{25.26} \right)^4 \right\}$$

$$= 1109 \text{ psi [7.65 MPa]}$$

in which

$$F_c = 1200 \text{ psi and } E = 1,700,000 \text{ (from Table 5-3)}$$

and

$$K = 0.671 \sqrt{\frac{E}{F_c}} = 0.671 \sqrt{\frac{1,700,000}{1200}} = 25.26$$

The required load is thus

$$A = \frac{\text{load}}{F'_c} = \frac{40,000}{1109} = 36.07 \text{ in.}^2 \; [23,272 \text{ mm}^2]$$

and the choices remain the same as for the 4-ft column.

If the 6-in. nominal thickness is used for the 16-ft column, we determine that

$$\frac{L}{d} = \frac{16 \times 12}{5.5} = 34.9$$

Since this is greater than the value of K, the stress condition is that of Zone 3 (Figure 13-2), and the allowable stress is

$$F'_c = \frac{0.30E}{(L/d)^2} = \frac{(0.30)(1,700,000)}{34.9^2} = 419 \text{ psi } [2.89 \text{ MPa}]$$

which requires an area for the column of

$$A = \frac{\text{load}}{F'_c} = \frac{40,000}{419} = 95.5 \text{ in.}^2 \; [61,617 \text{ mm}^2]$$

This is greater than the area for the largest section with a nominal thickness of 6 in., as listed in Table 6-7. Although larger sections may be available in some areas, it is highly questionable to use a member with these proportions as a column. Therefore, we consider the next larger nominal thickness of 8 in. Then, if

$$\frac{L}{d} = \frac{16 \times 12}{7.5} = 25.6$$

we are still in the Zone 3 condition, and the allowable stress is

$$F'_c = \frac{0.30E}{(L/d)^2} = \frac{(0.30)(1,700,000)}{25.6^2} = 778 \text{ psi } [5.36 \text{ MPa}]$$

which requires an area of

$$A = \frac{\text{load}}{F'_c} = \frac{40,000}{778} = 51.4 \text{ in.}^2 \; [33,163 \text{ mm}^2]$$

The smallest member usable is thus an 8 × 8. It is interesting to note that the required square column remains the same for all the column lengths, even though the stress varies from 1200 psi to 778 psi.

This is not uncommon and is simply due to the limited number of sizes available for the square column section.

Note: For the following problems use Douglas fir–larch, No. 1 grade.

Problems 13-2-A*-B-C-D. Find the allowable axial compression load for each of the following.

	Nominal Size (in.)	Unbraced Length (ft)	(m)
A	4 × 4	8	2.44
B	6 × 6	12	3.66
C	8 × 8	18	5.49
D	8 × 8	14	4.27

Problems 13-2-E *-F-G-H. Select the smallest square section for each of the following.

	Axial Load (kips)	(kN)	Unbraced Length (ft)	(m)
E	20	89	8	2.44
F	50	222	12	3.66
G	50	222	20	6.10
H	100	445	16	4.88

13-3. Steel Columns

Of the two major axes of a Standard I-beam, the moment of inertia about the axis parallel to the web is much the smaller; hence for the amount of material in the cross section, I-beams are not economical shapes when used as columns or struts. In former years built-up sections such as Figure 13-4c and 13-4d were used extensively, but wide flange sections (Figure 13-4a) are now rolled in a large variety of sizes and are used universally because they require a minimum of fabrication. They are sometimes called H-columns. For excessive loads or unusual conditions, plates are welded to the flanges of wide flange sections to give added strength (Figure 13-4b). The compression members of steel trusses are often formed of two angles, as shown in Figure 13-4e.

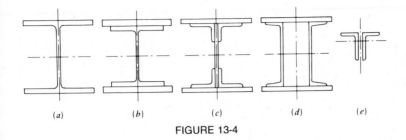

FIGURE 13-4

In the design of timber columns, the term *slenderness ratio* is defined as the unbraced length divided by the dimension of the least side, both in inches. For structural shapes such as those shown in Figure 13-4, the least lateral dimension is not an accurate criterion; and the radius of gyration r, which relates more precisely to the stiffness of columns in general, is used in steel column design. As discussed in Section 6-6, $r = \sqrt{I/A}$. For rolled sections the value of the radius of gyration with respect to both major axes is given in the tables of properties for designing. For built-up sections it may be necessary to compute its value. The slenderness ratio of a steel column is then l/r, where l is the effective length of the column in inches and r is the *least* radius of gyration of the cross section, also in inches. The slenderness ratio for compression members should not exceed 200.

The AISC Specification requires that, in addition to the unbraced length of a column, the condition of the ends must be given consideration. The slenderness ratio is Kl/r, where K is a factor dependent on the restraint at the ends of a column and the means available to resist lateral motion. Figure 13-5 shows diagrammatically six idealized conditions in which joint rotation and joint translation are illustrated. The term K is the ratio of the effective column length to the actual unbraced length. For average conditions in building construction, the value of K is taken as one; therefore the slenderness ratio Kl/r becomes simply l/r. (See Figure 13-5d).

The AISC Specification gives the following requirements for use in the design of compression members. The allowable unit stresses shall not exceed the following values:

1. On the gross section of axially loaded compression members, when Kl/r, the largest effective slenderness ratio of any unbraced segment is less than C_c,

	(a)	(b)	(c)	(d)	(e)	(f)
Buckled shape of column is shown by dashed line	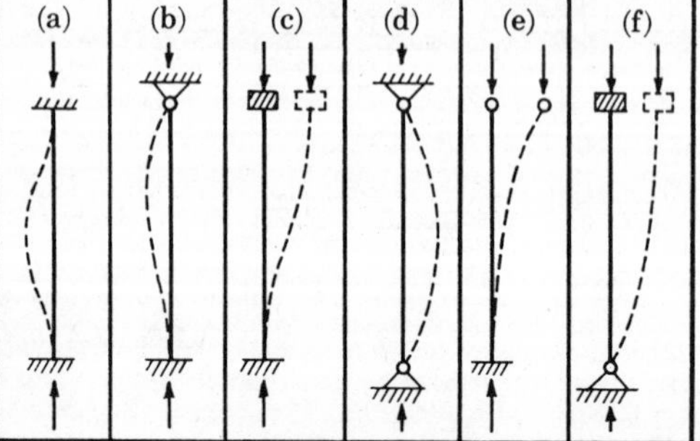					
Theoretical K value	0.5	0.7	1.0	1.0	2.0	2.0
Recommended design value when ideal conditions are approximated	0.65	0.80	1.2	1.0	2.10	2.0
End condition code	Rotation fixed and translation fixed					
	Rotation free and translation fixed					
	Rotation fixed and translation free					
	Rotation free and translation free					

FIGURE 13-5.   Determination of effective column length. (Reprinted from the *Manual of Steel Construction*, 8th ed., with permission of the publishers, American Institute of Steel Construction.)

$$F_a = \frac{[1 - (Kl/r)^2/2C_c^2] F_y}{\text{FS}} \quad \text{(Formula 13-3-1)}$$

where

$$\text{FS} = \text{factor of safety} = \frac{5}{3} + \frac{3(Kl/r)}{8C_c} - \frac{(Kl/r)^3}{8C_c^3}$$

and

$$C_c = \sqrt{\frac{2\pi^2 E}{F_y}}$$

2.   On the gross section of axially loaded columns, when Kl/r exceeds C_c,

$$F_a = \frac{12\pi^2 E}{23(Kl/r)^2} \qquad \text{(Formula 13-3-2)}$$

3. On the gross section of axially loaded bracing and secondary members, when l/r exceeds 120 (for this case, K is taken as unity),

$$F_{as} = \frac{F_a(\text{by Formula 13-3-1 or 13-3-2})}{1.6 - l/200r} \qquad \text{(Formula 13-3-3)}$$

In these formulas

F_a = axial compression stress permitted in the absence of bending stress

K = effective length factor (see Section 6-4)

l = actual unbraced length

r = governing radius of gyration (usually the least)

$C_c = \sqrt{2\pi^2 E/F_y}$; for A36 steel $C_c = 126.1$

F_y = minimum yield point of the steel being used (for A36 steel $F_y = 36,000$)

FS = factor of safety (see above)

E = modulus of elasticity of structural steel, 29,000 ksi,

F_{as} = axial compressive stress permitted in the absence of bending stress for bracing and other secondary members

To determine the allowable axial load that a main column will support, F_a, the allowable unit stress, is computed by Formula (1) or (2), and this stress is multiplied by the cross-sectional area of the column. If the column is a secondary member or is used for bracing, Formula (3) gives the allowable unit stress; these allowable unit stresses are somewhat greater than those permitted for main members. Table 13-1 gives allowable stresses computed in accordance with these formulas. It should be examined carefully because it will be of great assistance. Note particularly that this table is for use with A36 steel; tables based on other grades of steel are contained in the AISC Manual.

Example. A $W\,12 \times 65$ is used as a column with an unbraced length of 16 ft [4.88 m]. Compute the allowable load.

Solution: Referring to Table 6-1, we find that $A = 19.1$ in^2 [12,323 mm^2], $r_x = 5.28$ in. [134 mm], and $r_Y = 3.02$ in. [76.7 mm]. Because the column is unbraced with respect to both axes, the least radius of

Table 13-1. Allowable Unit Stresses for Columms of A36 Steel (ksi)

Main and Secondary Members Kl/r not over 120					
$\frac{Kl}{r}$	F_a (ksi)	$\frac{Kl}{r}$	F_a (ksi)	$\frac{Kl}{r}$	F_a (ksi)
1	21.56	41	19.11	81	15.24
2	21.52	42	19.03	82	15.13
3	21.48	43	18.95	83	15.02
4	21.44	44	18.86	84	14.90
5	21.39	45	18.78	85	14.79
6	21.35	46	18.70	86	14.67
7	21.30	47	18.61	87	14.56
8	21.25	48	18.53	88	14.44
9	21.21	49	18.44	89	14.32
10	21.16	50	18.35	90	14.20
11	21.10	51	18.26	91	14.09
12	21.05	52	18.17	92	13.97
13	21.00	53	18.08	93	13.84
14	20.95	54	17.99	94	13.72
15	20.89	55	17.90	95	13.60
16	20.83	56	17.81	96	13.48
17	20.78	57	17.71	97	13.35
18	20.72	58	17.62	98	13.23
19	20.66	59	17.53	99	13.10
20	20.60	60	17.43	100	12.98
21	20.54	61	17.33	101	12.85
22	20.48	62	17.24	102	12.72
23	20.41	63	17.14	103	12.59
24	20.35	64	17.04	104	12.47
25	20.28	65	16.94	105	12.33
26	20.22	66	16.84	106	12.20
27	20.15	67	16.74	107	12.07
28	20.08	68	16.64	108	11.94
29	20.01	69	16.53	109	11.81
30	19.94	70	16.43	110	11.67
31	19.87	71	16.33	111	11.54
32	19.80	72	16.22	112	11.40
33	19.73	73	16.12	113	11.26
34	19.65	74	16.01	114	11.13
35	19.58	75	15.90	115	10.99
36	19.50	76	15.79	116	10.85
37	19.42	77	15.69	117	10.71
38	19.35	78	15.58	118	10.57
39	19.27	79	15.47	119	10.43
40	19.19	80	15.36	120	10.28

Main Members Kl/r 121 to 200				Secondary Members [a] l/r 121 to 200			
$\frac{Kl}{r}$	F_a (ksi)	$\frac{Kl}{r}$	F_a (ksi)	$\frac{l}{r}$	F_{as} (ksi)	$\frac{l}{r}$	F_{as} (ksi)
121	10.14	161	5.76	121	10.19	161	7.25
122	9.99	162	5.69	122	10.09	162	7.20
123	9.85	163	5.62	123	10.00	163	7.16
124	9.70	164	5.55	124	9.90	164	7.12
125	9.55	165	5.49	125	9.80	165	7.08
126	9.41	166	5.42	126	9.70	166	7.04
127	9.26	167	5.35	127	9.59	167	7.00
128	9.11	168	5.29	128	9.49	168	6.96
129	8.97	169	5.23	129	9.40	169	6.93
130	8.84	170	5.17	130	9.30	170	6.89
131	8.70	171	5.11	131	9.21	171	6.85
132	8.57	172	5.05	132	9.12	172	6.82
133	8.44	173	4.99	133	9.03	173	6.79
134	8.32	174	4.93	134	8.94	174	6.76
135	8.19	175	4.88	135	8.86	175	6.73
136	8.07	176	4.82	136	8.78	176	6.70
137	7.96	177	4.77	137	8.70	177	6.67
138	7.84	178	4.71	138	8.62	178	6.64
139	7.73	179	4.66	139	8.54	179	6.61
140	7.62	180	4.61	140	8.47	180	6.58
141	7.51	181	4.56	141	8.39	181	6.56
142	7.41	182	4.51	142	8.32	182	6.53
143	7.30	183	4.46	143	8.25	183	6.51
144	7.20	184	4.41	144	8.18	184	6.49
145	7.10	185	4.36	145	8.12	185	6.46
146	7.01	186	4.32	146	8.05	186	6.44
147	6.91	187	4.27	147	7.99	187	6.42
148	6.82	188	4.23	148	7.93	188	6.40
149	6.73	189	4.18	149	7.87	189	6.38
150	6.64	190	4.14	150	7.81	190	6.36
151	6.55	191	4.09	151	7.75	191	6.35
152	6.46	192	4.05	152	7.69	192	6.33
153	6.38	193	4.01	153	7.64	193	6.31
154	6.30	194	3.97	154	7.59	194	6.30
155	6.22	195	3.93	155	7.53	195	6.28
156	6.14	196	3.89	156	7.48	196	6.27
157	6.06	197	3.85	157	7.43	197	6.26
158	5.98	198	3.81	158	7.39	198	6.24
159	5.91	199	3.77	159	7.34	199	6.23
160	5.83	200	3.73	160	7.29	200	6.22

[a] K taken as 1.0 for secondary members.
Source: Reprinted from the *Manual of Steel Construction*, 8th ed., with permission of the publishers, American Institute of Steel Construction.

gyration is used to determine the slenderness ratio. Also, with no qualifying conditions given, $K = 1.0$. The slenderness ratio is then

$$\frac{Kl}{r} = \frac{1 \times 16 \times 12}{3.02} = 63.6$$

$$\left[\frac{4.88 \times 10^3}{76.7} = 63.6 \right]$$

In design work it is usually considered acceptable to round the slenderness ratio off in front of the decimal point because a typical lack of
accuracy in the design data does not warrant greater precision. Therefore we consider the slenderness ratio to be 64; the allowable stress
given in Table 13-1 is $F_a = 17.04$ ksi [117.5 MPa]. The allowable
load on the column is then

$$P = A \times F_a = 19.1 \times 17.04 = 325.5 \text{ kips}$$

$$\left[P = \frac{12,323 \times 117.5}{10^3} = 1448 \text{ kN} \right]$$

Note: In the following problems assume A36 steel and a value of K
$= 1$.

Problem 13-3-A*. Compute the allowable axial load on a W 10×88 column with
an unbraced height of 15 ft [4.57 m].

Problem 13-3-B. A W 12×79 used as a column has an unbraced height of 22 ft
[6.71 m]. Compute the allowable axial load.

14

Connectors for Structural Steel

||||||||||||||||||||||||||||||||||||||||||||||

14-1. General

Structural steel members are most often connected by either welding or high-strength bolts. For bolting, flat elements are mated with matching holes in which the bolt acts as a pin. Holes reduce the cross-section area of the members—a special problem when the connection must resist tension.

14-2. Bearing-Type Shear Connections

The diagrams in Figure 14-1 show a simple connection between two steel bars that functions to transfer a tension force from one bar to another. Although this is a tension-transfer connection, it is also referred to as a shear connection because of the manner in which the connecting device (the bolt) works in the connection. (See Figure 14.1c.) If the bolt tension (owing to tightening of the nut) is relatively low, the bolt serves primarily as a pin in the matched holes, bearing against the sides of the holes as shown at Figure 14-1d. In addition to these functions, the bars develop tension stress that will be a maximum at the section through the bolt holes.

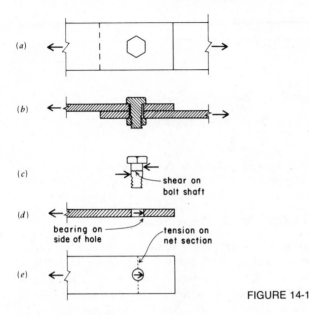

FIGURE 14-1

In the connection shown in Figure 14-1 the failure of the bolt in-volves a slicing (shear) failure that is developed as a shear stress on the bolt cross section. The resistance of the bolt can be expressed as an allowable shear stress F_v times the area of the bolt cross section, or

$$R = F_v \times A$$

With the size of the bolt and the grade of steel known it is a simple matter to establish this limit. In some types of connections it may be necessary to slice the same bolt more than once to separate the con-nected parts. This is the case in the connection shown in Figure 14-2, in which it may be observed that the bolt must be sliced twice to make the joint fail. When the bolt develops shear on only one section (Figure 14-1) it is said to be in *single shear*; when it develops shear on two sections (Figure 14-2) it is said to be in *double shear*.

When the bolt diameter is large or the bolt is made of strong steel the connected parts must be sufficiently thick if they are to develop the full capacity of the bolts. The maximum bearing stress permitted for this situation by the AISC Specification is $F_p = 1.5F_u$, and F_u is the

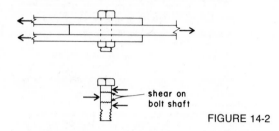

FIGURE 14-2

ultimate tensile strength of the steel in the part in which the hole occurs.

14-3. Friction-Type Shear Connections

When the nut on a bolt is sufficiently secure, the parts being bolted together will be squeezed so tightly that the connection will develop a friction stress between the two connected parts. If this friction is of sufficient magnitude, the bearing action described in Sect. 14-2 will not occur because the parts must slip somewhat to permit the bearing stress to develop. This type of joint is preferred in some cases when joint tightness or potential loosening is a problem.

High-strength bolts are those that function primarily in the friction-resistive manner just described. They are tightened to a carefully controlled degree so that a determined magnitude of the clamping effect for the development of friction can be considered reliable. If the friction resistance is depended on entirely, the bolt is called a friction-type fastener. If friction is used but the ultimate resistance is in bearing, it is called a bearing-type fastener. For the bearing-type fastener there are two different specifications for loading; one when the bolt threads are excluded from the shear failure planes, the other when they are not.

14-4. Tension Connections

When tension members have reduced cross sections two stress investigations must be considered. This is the case for members with holes for bolts or for bolts or rods with cut threads. For the member with a hole (Figure 14-1e) the allowable tension stress at the reduced cross section through the hole is $0.50F_u$, where F_u is the ultimate tensile strength of the steel. The total resistance at this reduced section (also called the net section) must be compared with the resistance at other, unreduced, sections at which the allowable stress is $0.60\,F_y$.

For threaded steel rods the maximum allowable tension stress at the threads is $0.33\,F_u$. For steel bolts the allowable stress is specified as a value based on the type of bolt. The load capacity of various types and sizes of bolt is given in Table 14-1.

When tension elements consist of W, M, S, and tee shapes the tension connection is usually not made in a manner that results in the attachment of all the parts of the section (e.g., both flanges plus the web for a W). In such cases the AISC Specification requires the determination of a reduced effective net area A_e that consists of

$$A_e = C_t A_n$$

where

A_n = the actual net area of the member
C_t = a reduction coefficient

Unless a larger coefficient can be justified by tests, the following values are specified:

1. For W, M, or S shapes with flange widths not less than two-thirds the depth and structural tees cut from such shapes, when the connection is to the flanges and has at least three fasteners per line in the direction of stress $-C_t = 0.90$.
2. For W, M, or S shapes not meeting the foregoing conditions and for tees cut from such shapes, provided the connection has not fewer than three fasteners per line in the direction of stress, $-C_t = 0.85$.
3. All members with connections that have only two fasteners per line in the direction of stress $-C_t = 0.75$.

Angles used as tension members are often connected by only one leg. In a conservative design the effective net area is only that of the connected leg, less the reduction caused by bolt holes. Rivet and bolt holes are punched larger in diameter than the nominal diameter of the fastener. The punching also damages a small amount of the steel around the perimeter of the hole; consequently the diameter of the hole to be

TABLE 14-1. Capacity of Structural Bolts (kips)

ASTM Designation	Connection Type[a]	Loading Condition[b]	Nominal Diameter (in.)							
			5/8	3/4	7/8	1	1-1/8	1-1/4	1-3/8	1-1/2
			Area, Based on Nominal Diameter (in²)							
			0.3068	0.4418	0.6013	0.7854	0.9940	1.227	1.485	1.767
A307		S	3.1	4.4	6.0	7.9	9.9	12.3	14.8	17.7
		D	6.1	8.8	12.0	15.7	19.9	24.5	29.7	35.3
		T	6.1	8.8	12.0	15.7	19.9	24.5	29.7	35.3
A325	F	S	5.4	7.7	10.5	13.7	17.4	21.5	26.0	30.9
		D	10.7	15.5	21.0	27.5	34.8	42.9	52.0	61.8
	N	S	6.4	9.3	12.6	16.5	20.9	25.8	31.2	37.1
		D	12.9	18.6	25.3	33.0	41.7	51.5	62.4	74.2
	X	S	9.2	13.3	18.0	23.6	29.8	36.8	44.5	53.0
		D	18.4	26.5	36.1	47.1	59.6	73.6	89.1	106.0
	All	T	13.5	19.4	26.5	34.6	43.7	54.0	65.3	77.7
A490	F	S	6.7	9.7	13.2	17.3	21.9	27.0	32.7	38.9
		D	13.5	19.4	26.5	34.6	43.7	54.0	65.3	77.7
	N	S	8.6	12.4	16.8	22.0	27.8	34.4	41.6	49.5
		D	17.2	24.7	33.7	44.0	55.7	68.7	83.2	99.0
	X	S	12.3	17.7	24.1	31.4	39.8	49.1	59.4	70.7
		D	24.5	35.3	48.1	62.8	79.5	98.2	119.0	141.0
	All	T	16.6	23.9	32.5	42.4	53.7	66.3	80.2	95.4

[a] F = friction; N = bearing, threads not excluded; X = bearing, threads excluded.

[b] S = single shear; D = double shear; T = tension.

Source: Reproduced from data in the Manual of Steel Construction, 8th ed., with permission of the publishers, American Institute of Steel Construction.

deducted in determining the net section is $\frac{1}{8}$ in. greater than the nominal diameter of the rivet.

14-5. Structural Bolts

Bolts used for the connection of structural steel members come in two types. Bolts designated A307 and called *unfinished* have the lowest load capacity of the structural bolts. The nuts for these bolts are tightened just enough to secure a snug fit of the attached parts; because of this, plus the oversizing of the holes, there is some movement in the development of full resistance. These bolts are generally not used for major connections, especially when joint movement or loosening under vibration or repeated loading may be a problem.

Bolts designated A325 or A490 are called *high-strength bolts*. The nuts of these bolts are tightened to produce a considerable tension force which results in a high degree of friction resistance between the attached parts. High-strength bolts are further designated as F, N, or X. The F designation denotes bolts for which the limiting resistance is that of friction. The N designation denotes bolts that function ultimately in bearing and shear but for which the threads are not excluded from the bolt shear planes. The X designation denotes bolts that function like the N bolts but for which the threads are excluded from the shear planes.

When bolts are loaded in tension their capacities are based on the development of the ultimate resistance in tension stress at the reduced section through the threads. When loaded in shear, bolt capacities are based on the development of shear stress in the bolt shaft. The shear capacity of a single bolt is further designated as S for single shear (Figure 14-1) or D for double shear (Figure 14-2). The capacities of structural bolts in both tension and shear are given in Table 14-1. The size range given in the table—$\frac{5}{8}$ in. to $1\frac{1}{2}$ in.—is that listed in the AISC Manual. However, the most commonly used sizes for structural steel framing are $\frac{3}{4}$ and $\frac{7}{8}$ in.

For a given diameter of bolt, there is a minimum thickness required for the bolted parts in order to develop the full shear capacity of the bolt. This thickness is based on the bearing stress between the bolt and the side of the hole, which is limited to a maximum of $F_p = 1.5F_u$. The stress limit may be established by either the bolt steel or the steel of the bolted parts.

14-6. Capacity of Bolted Connections

Use of the material developed in the preceding chapters is demonstrated in the following examples.

Example 1. Find the capacity of the connection shown in Figure 14-3 as limited by the load limit for the bolts and consideration of bearing and tension stresses in the steel plates. Bolts are $\frac{7}{8}$ in. [22 mm] A325F, and the steel of the plates has $F_y = 36$ ksi [250 MPa] and $F_u = 58$ ksi [400 MPa].

Solution: For the bolts the load condition is one of single shear, and the capacity of a single bolt is found in Table 14-1 as the S value of 10.5 kips [46.7 kN] per bolt. With two bolts the connection load limit is thus

$$T = 2 \times 10.5 = 21 \text{ k } [93.4 \text{ kN}]$$

For the plate in tension we first consider the capacity based on the gross (unreduced) section for which

$$A_g = 4 \times \tfrac{1}{2} = 2.0 \text{ in}^2. \ [1200 \text{ mm}^2]$$

$$T = 0.6 \, F_y A_g = 0.6 \times 36 \times 2.0 = 43.2 \text{ k } [180 \text{ kN}]$$

We next consider the net section through the hole, considering the

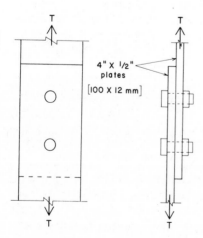

4" X 1/2"
plates

[100 X 12 mm]

FIGURE 14-3

hole to be $\frac{1}{8}$ in. larger than the bolt diameter. The net area at the section is thus

$$A_n = (4 - 1) \times \tfrac{1}{2} = 1.5 \text{ in}^2. \text{ [900 mm}^2]$$

As discussed in Section 14-1, the effective area of this connection must be reduced by a factor of 0.75, as the connection has only two bolts. Thus

$$A_e = 0.75 \, A_n = 0.75 \times 1.5 = 1.125 \text{ in}^2. \text{ [675 mm}^2]$$

Then

$$T = 0.50 \, F_u A_e = 0.50 \times 58 \times 1.125 = 32.625 \text{ k [135 kN]}$$

Finally, for bearing we consider the allowable bearing stress $F_p = 1.5 F_u$ to be developed by the area equal to the profile of the bolts, thus

$$A_b = \text{(number of bolts)} \times \text{(diameter of bolts)}$$

$$\times \text{(thickness of the plate)}$$

$$= 2 \times \tfrac{7}{8} \times \tfrac{1}{2} = 0.875 \text{ in}^2. \text{ [528 mm}^2]$$

and

$$T = F_p A_b = (1.5 \times 58) \times 0.875 = 76.125 \text{ k [317 kN]}$$

The least of the values determined for T is the limit for the connection. It may thus be observed that the connection is limited by the capacity of the bolts to a load of 21 k [93.4 kN].

Example 2.   Find the capacity of the connection shown in Figure 14-4 as limited by the load limit for the bolts and consideration of bearing and tension stresses in the steel plates. Bolts are $\frac{3}{4}$ in. [19 mm] A325F, and the steel of the plates has $F_y = 36$ ksi [250 MPa] and $F_u = 58$ ksi [400 MPa].

Solution: In this case the bolts act in double shear and the capacity of a single bolt is found in Table 14-1 as the D value of 15.5 kips [69 kN]. With six bolts the connection load limit is thus

$$T = 6 \times 15.5 = 93 \text{ k [414 kN]}$$

Bearing and tension must be investigated separately for the inner plate and for the two outer plates working in tandem. For the inner

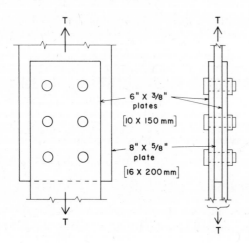

FIGURE 14-4

plate we find

$$A_g = 8 \times \tfrac{5}{8} = 5.0 \text{ in}^2. \text{ [3200 mm}^2\text{]}$$

$$A_n = (8 - 1.75) \times \tfrac{5}{8} = 3.91 \text{ in}^2. \text{ [2496 mm}^2\text{]}$$

(Note that deduction is made for two holes at the reduced section for the inner plate.)

$$A_b = 6 \times \tfrac{3}{4} \times \tfrac{5}{8} = 2.8125 \text{ in}^2. \text{ [1824 mm}^2\text{]}$$

We next find the corresponding values for the pair of outer plates, thus

$$A_g = 2 \times 6 \times \tfrac{3}{8} = 4.5 \text{ in}^2. \text{ [3000 mm}^2\text{]}$$

$$A_n = 2 \times (6 - 1.75) \times \tfrac{3}{8} = 3.1875 \text{ in}^2. \text{ [2120 mm}^2\text{]}$$

$$A_b = 2 \times 6 \times \tfrac{3}{4} \times \tfrac{3}{8} = 3.375 \text{ in}^2. \text{ [2280 mm}^2\text{]}$$

Comparing these areas, we observe the following:

1. Stress in tension on the gross area is critical in the outer plates and

$$T = 0.6 F_y A_g = 0.6 \times 36 \times 4.5 = 97.2 \text{ k [450 kN]}$$

2. Stress in tension on the net area is critical in outer plates and

$$T = 0.5 F_u A_n = 0.5 \times 58 \times 3.1875 = 92.44 \text{ k [424 kN]}$$

3. Stress in bearing is critical on the inner plate and

$$T = F_p A_b = 1.5 \times 58 \times 2.8125 = 245 \text{ k [1094 kN]}$$

The connection is thus seen to be limited by the tension stress on the net area of the outer plates to a load of 92.44 k, which is just slightly less than the capacity of the six bolts. (Note that the limit in SI units is the capacity of the bolts at 414 kN.)

14-6-A*. A connection employing steel bolts is similar to that shown in Figure 14-3 with the following data. Bolts are $\frac{7}{8}$-in. diameter type A325F, and there are three bolts in a single line. Plates are 5 in. [125 mm] wide and $\frac{1}{4}$ in. [6 mm] in thickness and are made from steel with $F_y = 36$ ksi [250 MPa] and $F_u = 58$ ksi [400 MPa]. Find the capacity of the connection as limited by the load limit for the bolts and consideration of bearing and tension stresses in the plates.

Problem 14-6-B. A connection employing steel bolts is similar to that shown in Figure 14-4 with the following data. Bolts are 1-in. diameter [25 mm] type A325F, and there are four bolts arranged in two rows of two. Outer plates are 7 in. [175 mm] wide and $\frac{1}{2}$ in. [12 mm] thick. The inner plate is 9 in. [225 mm] wide and $\frac{3}{4}$ in. [19 mm] thick. All plates have steel with $F_y = 36$ ksi [250 MPa] and $F_u = 58$ ksi [400 MPa]. Find the capacity of the connection as limited by the load limit for the bolts and consideration of bearing and tension stresses in the plates.

14.7. Welding

One of the distinguishing characteristics of welded construction is the facility with which one member may be attached directly to another without the use of additional plates or angles, which are necessary in bolted and riveted connections. A welded connection requires no holes for fasteners; therefore the gross rather than the net section may be considered when determining the effective cross-sectional area of members in tension.

Moment-resisting connections are readily achieved by welding; consequently welded connections are customary in construction in order to develop continuity in the framing. Welding may also be used in noncontinuous construction, but care must be exercised in design to ensure that a rigid connection is not provided where free-end conditions have been assumed in the design of the framing.

Welding is often used in combination with bolting in "shop-welded and field-bolted construction." Here connection angles with holes in the outstanding legs may be welded to a beam in the fabricating shop and then bolted to a girder or column in the field.

Although there are many welding processes, electric arc welding is the one generally used in steel building construction. In this type of welding an electric arc is formed between an electrode and the two pieces of metal that are to be joined. The intense heat melts a small portion of the members to be joined as well as the end of the electrode or metallic wire. The term *penetration* is used to indicate the depth from the original surface of the base metal to the point at which fusion ceases. The globules of melted metal from the electrode flow into the molten seat and, when cool, are united with the members that are to be welded together. *Partial penetration* is the failure of the weld metal and base metal to fuse at the root of a weld. It may result from a number of items, and such incomplete fusion produces welds that are inferior to those of full penetration.

14-8. Welded Joints

The weld most commonly used for structural steel in building construction is the *fillet weld*. It is approximately triangular in cross section and is formed between the two intersecting surfaces of the joined members. See Figure 14-5*a* and 14-5*b*. The *size* of a fillet weld is the leg length of the largest inscribed isosceles right triangle, *AB* or *BC*. (See Figure 14-5*a*.) The *root* of the weld is the point at the bottom of the weld, point *B* in Figure 14-5*a*. The *throat* of a fillet weld is the distance from the root to the hypotenuse of the largest isosceles right triangle that can be inscribed within the weld cross section, distance *BD* in Figure 14-5*a*. The exposed surface of a weld is not the plane surface indicated in Figure 14-5*a* but is usually somewhat convex, as shown in Figure 14-5*b*. Therefore the actual throat may be greater than that shown in Figure 14-5*a*. This additional material is called *reinforcement*. It is not included in determining the strength of a weld.

A single-vee groove weld between two members of unequal thick-

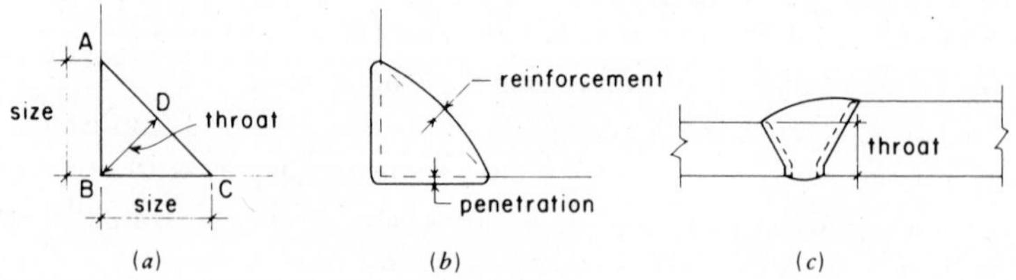

FIGURE 14-5

ness is shown in Figure 14-5c. The *size* of a butt weld is the thickness of the thinner part joined, with no allowance made for the weld reinforcement.

If the dimension (size) of *AB* in Figure 14-5a is 1 unit in length, $(AD)^2 + (BD)^2 = 1^2$. Because *AD* and *BD* are equal, $2(BD)^2 = 1^2$ and $BD = \sqrt{0.5}$ or 0.707. Therefore the throat of a fillet weld is equal to the *size* of the weld multiplied by 0.707. As an example, consider a $\frac{1}{2}$-in. fillet weld. This would be a weld with dimensions *AB* or *BC* equal to $\frac{1}{2}$ in. In accordance with the foregoing, the throat would be 0.5 × 0.707 or 0.3535 in. Then, if the allowable unit shearing stress on the throat is 21 ksi, the allowable working strength of a $\frac{1}{2}$-in. fillet weld is $0.3535 \times 21 = 7.42$ kips *per linear inch of weld*. If the allowable unit stress is 18 ksi, the allowable working strength is $0.3535 \times 18 = 6.36$ kips *per linear inch of weld*.

The permissible unit stresses used in the preceding paragraph are for welds made with E 70 XX and E 60 XX type electrodes on A36 Steel. Particular attention is called to the fact that *the stress in a fillet weld is considered as shear on the throat, regardless of the direction of the applied load*. Neither plug nor slot welds shall be assigned any values in resistances other than shear. The allowable working strengths of fillet welds of various sizes are given in Table 14-2 with values rounded to 0.1 of a kip.

TABLE 14-2. Allowable Working Strength of Fillet Welds

Size of Weld (in.)	Allowable Load (kips/in.)		Allowable Load (kN/mm)		Size of Weld (mm)
	E 60 XX Electrodes $F_{vw} = 18$ (ksi)	E 70 XX Electrodes $F_{vw} = 21$ (ksi)	E 60 XX Electrodes $F_{vw} = 124$ (MPa)	E 70 XX Electrodes $F_{vw} = 145$ (MPa)	
$\frac{3}{16}$	2.4	2.8	0.42	0.49	4.76
$\frac{1}{4}$	3.2	3.7	0.56	0.65	6.35
$\frac{5}{16}$	4.0	4.6	0.70	0.81	7.94
$\frac{3}{8}$	4.8	5.6	0.84	0.98	9.52
$\frac{1}{2}$	6.4	7.4	1.12	1.30	12.7
$\frac{5}{8}$	8.0	9.3	1.40	1.63	15.9
$\frac{3}{4}$	9.5	11.1	1.66	1.94	19.1

14-9. Design of Welded Joints

The most economical choice of weld to use for a given condition depends on several factors. It should be borne in mind that members to be connected by welding must be firmly clamped or held rigidly in position during the welding process. When riveting a beam to a column it is necessary to provide a seat angle as a support to keep the beam in position for riveting the connecting angles. The seat angle is not considered as adding strength to the connection. Similarly, seat angles are commonly used with welded connections. The designer must have in mind the actual conditions during erection and must provide for economy and ease in working the welds. Seat angles or similar members used to facilitate erection are *shop welded* before the material is sent to the site. The welding done during erection is called *field welding*. In preparing welding details the designer indicates shop or field welds on the drawings. Conventional welding symbols are used to identify the type, size, and position of the various welds. Only engineers or architects experienced in the design of welded connections should design or supervise welded construction. It is apparent that a wide variety of connections is possible; experience is the best aid in determining the most economical and practical connection.

The following examples illustrate the basic principles on which welded connections are designed:

Example 1. A bar of A36 steel, $3 \times \frac{7}{16}$ in. [76.2 × 11 mm] in cross section is to be welded with E 70 XX electrodes to the back of a channel so that the full tensile strength of the bar may be developed. What is the size of the weld? (See Figure 14-6.)

Solution: The area of the bar is $3 \times 0.4375 = 1.313$ in.^{2} [76.2 × 11 = 838.2 mm^2]. Because the allowable unit tensile stress of the steel

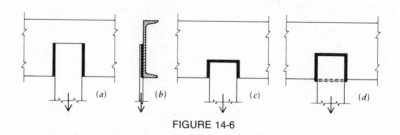

FIGURE 14-6

is 22 ksi (Table 5-2), the tensile strength of the bar is $F_t \times A = 22 \times 1.313 = 28.9$ kips [$152 \times 838.2/10^3 = 127$ kN]. The weld must be of ample dimensions to resist a force of this magnitude.

A $\frac{3}{8}$-in. [9.52 mm] fillet weld will be used. Table 14-2 gives the allowable working strength as 5.6 kips/in. [0.98 kN/mm]. Hence the required length of weld to develop the strength of the bar is 28.9 ÷ 5.6 = 5.16 in. [127 ÷ 0.98 = 130 mm]. The position of the weld with respect to the bar has several options, three of which are shown in Figure 14-6a, c, and d.

Example 2. A $3\frac{1}{2} \times 3\frac{1}{2} \times \frac{5}{16}$-in. [89 × 89 × 7.94 mm] angle of A36 steel subjected to a tensile load is to be connected to a plate by fillet welds using E 70 XX electrodes. What should the dimensions of the welds be to develop the full tensile strength of the angle?

Solution: We shall use a $\frac{1}{4}$-in. fillet weld which has an allowable working strength of 3.7 kips/in. [0.65 kN/mm] (Table 14-2). From Table 6-4 the cross-sectional area of the angle is 2.09 in.^{2} [1348 mm^2]. By using the allowable tension stress of 22 ksi [152 MPa] for A36 steel (Table 5-2), the tensile strength of the angle is 22 × 2.09 = 46 kips [$152 \times 1348/10^3 = 205$ kN]. Therefore the required total length of weld to develop the full strength of the angle is 46 ÷ 3.7 = 12.4 in. [205 ÷ 0.65 = 315 mm].

An angle is an unsymmetrical cross section, and the welds marked L_1 and L_2 in Figure 14-7 are made unequal in length so that their individual resistance will be proportioned in accordance to the distributed area of the angle. From Table 6-4 we find that the centroid of the angle section is 0.99 in. [25 mm] from the back of the angle; hence the two welds are 0.99 in. [25 mm] and 2.51 in. [64 mm] from the centroidal axis, as shown in Figure 14-7. The lengths of welds L_1 and L_2 are made inversely proportional to their distances from the axis, but the sum of

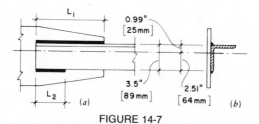

FIGURE 14-7

their lengths is 12.4 in. [315 mm]. Therefore

$$L_1 = \frac{2.51}{3.5} \times 12.4 = 8.9 \text{ in.}$$

$$\left[\frac{64}{89} \times 315 = 227 \text{ mm} \right]$$

and

$$L_2 = \frac{0.99}{3.5} \times 12.4 = 3.5 \text{ in.}$$

$$\left[\frac{25}{89} \times 315 = 88 \text{ mm} \right]$$

These are the design lengths required; for reinforcement each weld would actually be made $\frac{1}{4}$-in. [6.5 mm] longer than its computed length.

When angle shapes are used as tension members and connected by fastening only one leg, it is questionable to assume a stress distribution of equal magnitude on the entire cross section. Some designers therefore prefer to ignore the stress in the unconnected leg and to limit the capacity of the member in tension to the force obtained by multiplying the allowable stress by the area of the connected leg only. If this is done, it is logical to use welds of equal length on each side of the leg, as in Example 1.

Problem 14-9-A*. A 4 × 4 × $\frac{1}{2}$-in. angle of A36 steel is to be welded to a plate with E 70 XX electrodes to develop the full tensile strength of the angle. Using $\frac{3}{8}$-in. fillet welds, compute the design lengths L_1 and L_2, as shown in Figure 14-7, assuming the development of tension on the entire cross section of the angle.

Problem 14-9-B. Redesign the welded connection in Problem 14-9-A assuming that the tension force is developed only by the connected leg of the angle. (*Hint:* The connected leg of the angle behaves in the manner of a flat bar if the other leg of the angle is ignored.)

15

Torsional Effects

||||||||||||||||||||||||||||||||||||||||||||||||||

15-1. Torison

When a bar is firmly secured at one end and a force is applied to the other end so that the bar tends to twist, the stresses developed in the bar are *torsional stresses*. Figure 15-1a and b represents a bar subjected to two equal forces P that cause torsional stresses. Each force is a distance from the center of the bar, and the *twisting moment* or *torque* has a magnitude of 2Pa. It is important to realize that the forces do not *bend* the bar; they tend to twist it.

At any section of the bar, between the left (fixed) end and the section at which the loads are applied, the bar tends to shear off from the section adjacent to it. Figure 15-1c represents a portion of a bar or shaft that is subjected to torsional stresses, the left end being fixed. Before torsional stresses are applied, AB is a straight line on the surface of the cylinder parallel to a longitudinal axis through its center of gravity. When the shaft is subjected to torque, both stresses and deformation occur. The line AB takes the position of AC. If the maximum stress, the stress on the outermost fiber of the shaft, remains within the elastic limit of the material, the radial line OC remains straight. The unit shearing stress at any point in the shaft is directly proportional to its distance from the axis O.

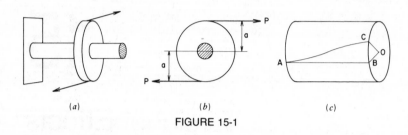

FIGURE 15-1

15-2.   Torsion Formula

Torque or twisting moment is the product of the resultant force and its distance from the axis about which it tends to turn. In Figure 15-1a and b there are two forces of P pounds each, and the moment arm of each force is a. If a is in units of inches, T, the torque, is $2Pa$ in-lb. For equilibrium T the torque or twisting moment must equal T_r, the resisting moment. The torsion formula for round shafts is

$$T = \frac{f_s J}{c}$$

in which

$T =$ the torque or twisting moment in inch-pounds

$f_s =$ the unit shearing stress on the fibers at the surface of the shaft in pounds per square inch

$J =$ the polar moment of inertia of the cross section of the shaft in inches to the fourth power

$c =$ the distance of the outermost fiber of the shaft (the radius) from the axis in inches

J, the *polar moment of inertia* of a surface about an axis through its centroid perpendicular to the surface, is equal to the sum of the products of all the elementary areas of the surface multiplied by the squares of their distances from the axis. For a circular area of which *D* is the diameter, $J = \pi D^4/32$.

For a hollow circular area, the cross-sectional area of a pipe or hollow shaft, in which *D* is the outside diameter and D_1 is the inside diameter,

$$J = \frac{\pi(D^4 - D_1^4)}{32}$$

The torsion formula is $T = f_s J/c$. Thus for a solid round shaft $c = D/2$ and $\pi = 3.1416$ and

$$T = \frac{f_s \times (3.1416 \times D^4/32)}{D/2} \text{ or } T = 0.196 f_s D^3$$

Similarly, for a hollow shaft,

$$T = \frac{0.196 f_s (D^4 - D_1^4)}{D}$$

In using these torsion formulas, the unit stress must not exceed the elastic limit of the material; T must be in units of inch-pounds and D in inches.

Example 1. A pulley wheel on a solid shaft has a diameter of 2 ft [600 mm], and the resultant load on the wheel is 1200 lb [5.34 kN]. If the shaft is 3 in. [75 mm] in diameter, compute the maximum unit shearing stress in the shaft.

Solution: The diameter of the wheel is 2 ft; therefore the moment arm of the 1200 lb force is 1 ft, or 12 in. [300 mm]. Thus $T = 1200 \times 12 = 14,400$ in.-lb [1602 kN-m]. Then

$$T = 0.196 f_s D^3$$

$$f_s = \frac{T}{0.196 \times D^3} = \frac{14,400}{0.196 \times (3)^3}$$

$$f_s = 2720 \text{ psi } [19.37 \text{ MPa}]$$

Example 2. The resultant load on a pulley wheel 20 in. [500 mm] in diameter on a solid shaft is 10,000 lb [45 kN]. If the maximum unit shearing stress is 8000 psi [55 MPa] for the shaft, what is the required diameter of the shaft?

Solution: The pulley wheel being 20 in. in diameter, the moment arm of the load is 10 in.; therefore, the torque is $10,000 \times 10 = 100,000$ in.-lb. Then

$$T = 0.196 \times f_s \times D^3$$

$$D^3 = \frac{T}{0.196 \times f_s} = \frac{100,000}{0.196 \times 8000} = 63.8$$

$$D = 4 \text{ in. } [101 \text{ mm}]$$

Example 3. A hollow shaft has outside and inside diameters of 4 in. and 2 in. [100, 50 mm], respectively. A load of 9000 lb [40 kN] on a pulley wheel on the shaft has a moment arm of 10 in. [250 mm]. Compute the maximum unit shearing stress in the shaft.

Solution: The torque $T = 9000 \times 10 = 90,000$ in.-lb [10 kN-m]. Then

$$T = \frac{0.196 \times f_s(D^4 - D_1^4)}{D}$$

$$90,000 = \frac{0.196 \times f_s(4^4 - 2^4)}{4} \quad \text{and} \quad f_s = 7653 \text{ psi } [54.4 \text{ MPa}]$$

Problem 15-2-A. A load of 19,600 lb [87 kN] is applied to a pulley wheel 20 in. [500 mm] in diameter attached to a solid shaft. Compute the diameter of the shaft if the maximum shearing unit stress is 8000 psi [55 MPa].

Problem 15-2-B*. The load on a pulley wheel 2 ft [600 mm] in diameter on a 4-in. [100 mm] solid shaft is 4000 lb [17.8 kN]. Compute the maximum shearing unit stress in the shaft.

Problem 15-2-C. A hollow shaft has an outside diameter of 3 in. [75 mm] and an inside diameter of 1 in. [25 mm]. Attached to the shaft is a pulley wheel 22 in. [550 mm] in diameter to which a 3800 lb [17 kN] load is applied. Compute the maximum shearing unit stress in the shaft.

15-3. Shaft Couplings

In the design of shafting, frequently it is necessary to join two pieces of shafting end to end. Shafts may be forged with flanges that serve as couplings, or the couplings may be made separately and attached to the shafts. In any event, the coupling must be strong enough to transfer the torque from one piece of shafting to the other. The flanges or faces of couplings are bolted together by a number of bolts arranged in a

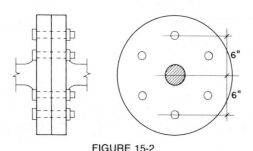

FIGURE 15-2

circle. These bolts tend to shear at the plane of contact of the couplings.

Example.   The ends of two shafts are joined by the coupling shown in Figure 15-2. A torque of 160,000 in.-lb [18 kN-m] is transmitted from one shaft to the other by six bolts $\frac{3}{4}$ in. [19 mm] in diameter arranged in a circle 12 in. [300 mm] in diameter. Compute the unit shearing stress in the bolts.

Solution: Let *P* be the force transmitted by each bolt. Then 6*P* is the force transmitted by all six bolts. Since the circle of bolts has a radius of 6 in., the torque exerted by all the bolts is $(6 \times 6 \times P)$ in.-lb. This torque must equal 160,000 in.-lb; therefore, $6 \times 6 \times P = 160,000$, and $P = 4444$ lb [20 kN].

The cross-sectional area of a single bolt is 0.4418 in.^{2} [284 × 10^3 mm^2], making the unit stress in the bolts 4440/0.4418 = 10,050 psi [70 MPa].

Problem 15-3-A.   A shaft coupling (Figure 15-2) contains a circle 16 in. [400 mm] in diameter on which $\frac{7}{8}$-in. [22 mm] bolts will be placed. If the torque transferred from one piece of shafting to the other is 200 kip-in. [23 kN-m] and the allowable unit shearing stress in the bolts is 8 ksi [55 MPa], how may bolts are required?

16

Stresses in Pipes and Tanks

IIIIIIIIIIIIIIIIIIIIIIIIIIIIIIIIIIIIIIIIIIIIII

16-1. General

When pressure from water, gas, or steam is exerted internally in a thin cylindrical container, such as a pipe or tank, tensile stresses are set up. A cylinder or pipe is considered "thin" when the thickness of the wall is small in comparison with the diameter. For such containers it is assumed that the tensile stresses are distributed uniformly over the thickness of the wall. These stresses must, of course, fall within the allowable tensile stress of the material of which the pipe or tank is made. Figure 16-1a and b shows a short section of a pipe in elevation and plan. The length of the pipe is l, the inside diameter is d, and the thickness of the pipe material is t. The radial arrows in Figure 16-1b show the direction of the steam or water pressure; the pressure is normal to the internal surface of the pipe.

Internal pressure tends to produce two different types of failure. In one of these the pipe or tank may fail in tension by a longitudinal opening parallel to the longitudinal axis, possibly along a seam. The other type is a tension failure along a line parallel to the circumference of a closed pipe or tank. The stresses tending to cause both types of

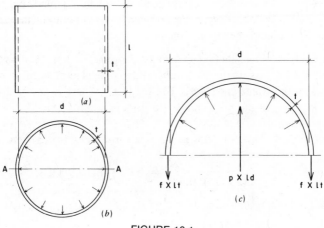

FIGURE 16-1

failure are tensile, but they are different in magnitude for the same internal pressure.

16-2. Circumferential Tension

Imagine that the pipe shown in Figure 16-1*a* and *b* is cut in half at the section marked $A - A$. The total internal pressure on the half pipe is equal to the pressure per unit of area *p* multiplied by a rectangular area, the dimensions of which are the length *l* and diameter *d*. Thus the total pressure on the half pipe is $p \times ld$; its resultant is shown by the arrow at the center of the pipe (Figure 16-1*c*), although it is distributed over the curved inner surface. This resultant force, $P \times ld$, must be held in equilibrium by the stresses in the two cut sections of the pipe as shown in the figure. Let *f* be the unit tensile stress in the pipe metal, and, since $l \times t$ is the area of pipe resisting tension, $f \times lt$ is the stress at each side, as shown in the figure. Then $(f \times lt) + (f \times lt) = p \times ld$, or

$$2flt = pld \quad \text{and} \quad f = \frac{pd}{2t}$$

Steam and water pressures are generally given in pounds per square inch; therefore, in the foregoing formula, d and t are in inches and f is in units of pounds per square inch.

Example.   The inside diameter of a steel pipe is 12 in. [300 mm] and its shell thickness is $\frac{3}{8}$ in. [9.5 mm]. If this pipe is used as a water main in which the water pressure is 100 psi [0.690 MPa], what is the unit tensile stress in the pipe parallel to its longitudinal axis (the circumferential unit tensile stress)?

Solution: By data p = 100 psi, d = 12 in., and t = 0.375 in. Then

$$f = \frac{pd}{2t} = \frac{100 \times 12}{2 \times 0.375} = 1600 \text{ psi } [10.9 \text{ MPa}]$$

which is the unit tensile stress.

16-3. Transverse Tension

In the preceding section we discussed circumferential tension in the wall (shell) of a pipe or cylindrical tank; this is sometimes called "hoop tension." The transverse tension to be considered here is the stress developed at right angles to the cross section, along a line parallel to the circumference of a closed cylinder. If the container shown in Figure 16-1a and b is a closed tank in which there is steam or water pressure, this pressure tends to force the head off the tank in addition to producing circumferential tension.

The area of the head is $\pi d^2/4$ and, if the internal pressure is p pounds per square inch, the total pressure on the head is $p \times (\pi d^2/4)$. The area of metal resisting this force is approximately the circumference of the container multiplied by the thickness of the metal, $\pi d \times t$. Now, if the tensile stress in the metals is f pounds per square inch, $f \times \pi dt$ is the total tensile force in the cross section of the pipe or tank, and therefore

$$f\pi dt = \frac{p\pi d^2}{4} \quad \text{and} \quad f = \frac{pd}{4t}$$

Example.   A capped steel pipe has an inside diameter of 12 in. [300 mm] and a wall thickness of $\frac{3}{8}$ in. [9.5 mm]. If the pressure in the pipe is 100 psi [0.690 MPa], what is the unit tensile stress in the metal on a section corresponding to a joint on the circumference?

Solution:

$$f = \frac{pd}{4t} = \frac{100 \times 12}{4 \times 0.375} = 800 \text{ psi [5.45 MPa]}$$

Note that this stress is one-half the unit stress in the pipe on a section parallel to the longitudinal axis (Sect. 16-2), although the pipe size and internal pressure are the same.

Problem 16-3-A.   A steel pipe has an inside diameter of 14 in. [350 mm], a wall thickness of 0.33 in. [8.4 mm], and is subjected to a water pressure of 80 psi. [0.550 MPa]. Compute (1) the circumferential unit tensile stress and (2) the transverse unit tensile stress in the pipe.

17

Reinforced Concrete

||||||||||||||||||||||||||||||||||||||||||||||||

17-1. Concrete and Steel in Combination

A reinforced concrete beam is a structural member composed of concrete in which steel bars are embedded. Whereas concrete in its hardened state is well able to resist compression, it is relatively weak in tension. To overcome this lack of tensile resistance, steel bars are placed in the concrete at the proper positions. In a simple beam tensile stress occurs in the lower portion of the beam where the bending moment is positive; consequently the steel reinforcing bars are placed near the bottom. For continuous beams tension is also developed in the upper portion over the supports where there is negative bending moment (Figure 9-7); here reinforcing bars are placed near the upper surface of the beam. In structural design it is assumed that the steel reinforcement resists all of the tension in a beam, the low tensile strength of the concrete being neglected.

17-2. Distribution of the Compressive Stress

For beams of symmetrical cross section composed of one material such as steel or wood (homogeneous beams), the neutral axis lies at mid-depth of the cross section. In a reinforced concrete beam the position of the neutral axis must be computed; that is, the distance identified as c in Figure 17-1 must be determined. Deferring consideration of the

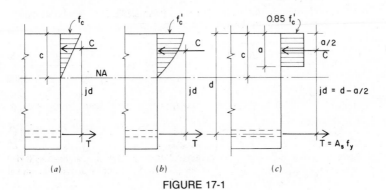

FIGURE 17-1

way in which this is accomplished, let us again direct our attention to Figure 17-1 which shows three different assumed distributions of the compressive stress. The diagram at (*a*) illustrates straight-line distribution in accordance with elastic theory. The stress varies directly as the distance from the neutral axis, at which it is zero, and increases to a maximum value at the compression face of the beam. This value is called the *extreme fiber stress* and is denoted by the symbol f_c. Figure 17-1*b* illustrates a parabolic distribution when the value of the extreme fiber stress has reached f'_c, the *specified compressive strength* (ultimate strength) of the concrete. This corresponds to inelastic behavior of concrete with an assumed parabolic stress-strain diagram. Figure 17-1*c* shows a rectangular compressive stress distribution which is assumed to be equivalent in its static effect to the parabolic pattern. The rectangular "stress block" is based on the assumption that a concrete stress of $0.85f'_c$ is uniformly distributed over a part of the compression zone with dimensions equal to the beam width *b* and the distance *a* which locates a line parallel to and above the neutral axis. This is the stress distribution most often used in design under the ultimate strength theory.

17-3. Design Methods

The design of reinforced concrete structural members may be accomplished by two different methods. One, called *working stress design* (sometimes designated WSD), is based on the straight-line distribution

TABLE 17-1. Design Values for Concrete Beams: Working Stress Method[a]

	Values of f'_c (psi)			
	2000	2500	3000	4000
$f_c = 0.45f'_c$ (psi)	900	1125	1350	1800
Modular ratio: $n = E_s/E_c$	11.3	10.1	9.2	8.0

[a] Adapted from data in *Building Code Requirements for Reinforced Concrete* (ACI 318-83) with permission of the publisher, American Concrete Institute. Values are for normal weight concrete, 145 lb/ft^3.

of compressive stress in the concrete (Figure 17-1a); the other is known as *ultimate strength design* (USD) and is now the predominant design method used for important building structures.

In working stress design a maximum allowable (working) value for the extreme fiber stress is established (Table 17-1), and the formulas are predicated on elastic behavior of the reinforced concrete member under service load. The straight-line distribution of compressive stress is valid at working stress levels because the stresses developed vary approximately with the distance from the neutral axis, in accordance with elastic theory. Shrinkage and cracking of the concrete, however, together with the phenomenon of creep under sustained loading, complicate the stress distribution. Over time, stresses computed in reinforced concrete members on the basis of elastic theory are not realistic. Generally speaking, the acceptable safety of the working stress design method is maintained by the differentials provided between the allowable compressive stress f_c and the specified compressive strength of the concrete f'_c and between the allowable tensile stress f_s and the yield strength of the steel reinforcement f_y. These, in effect, are measures of safety. The designation *strength design method* has been adopted by the American Concrete Institute to replace USD and is used hereafter in this text (sometimes abbreviated SDM).

Extended discussion of operational design procedures under either WSD or SDM is beyond the scope of this book. The material presented here is intended to serve as a basic introduction to structural design of reinforced concrete. For a more detailed treatment of design proce-

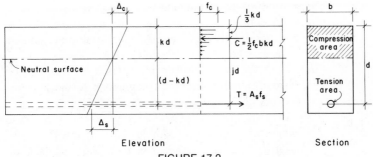

Elevation   Section

FIGURE 17-2

dures see *Simplified Design of Reinforced Concrete* by Harry Parker, Fifth Edition (New York: Wiley, 1984). This chapter emphasizes the strength design method but also includes discussion of flexural design of beams under the working stress design method.

17-4. Flexure Notation: Working Stress Design

The following notation is used in developing the WSD flexure formulas. The various symbols are identified in Figure 17-2, which is an amplification of Figure 17-1a except that the distance from the top of the beam to the neutral axis is called *kd* instead of *c*.

Δ_c and Δ_s = deformations per unit of length of the concrete and steel, respectively

E_s = modulus of elasticity of steel, in pounds per square inch; 29,000,000 psi

E_c = modulus of elasticity of concrete in compression, in pounds per square inch, the magnitude depending on the quality and weight of the concrete

n = ratio of modulus of elasticity of steel to that of concrete E_s/E_c

f_c = compressive unit stress on the concrete at the surface most remote from the neutral surface, in pounds per square inch

f_s = tensile unit stress in the longitudinal reinforcement, in pounds per square inch

b = width of the rectangular beam, in inches

d = depth from the compression face of the beam to the center of the longitudinal steel reinforcement, in inches: the *effective depth*

k = ratio of distance of the neutral axis of the cross section from extreme fibers in compression to the effective depth of the beam

kd = distance from the neutral axis of the cross section to the extreme fibers in compression, in inches

j = ratio of the distance between the resultant of the compressive stresses and center of tensile stresses to d, the effective depth of the beam

jd = distance between the resultant of the compressive stresses and the center of the tensile stresses: the lever arm of the resisting couple, in inches

A_s = area of the cross section of the longitudinal steel reinforcement, in square inches

p = ratio of the area of the cross section of the longitudinal steel reinforcement to the effective area of the concrete beam, $p = A_s/bd$. [The 1983 ACI Code replaces p with ρ (rho) as the symbol for reinforcement ratio]

M_c = resisting moment of the compressive stresses in the concrete, in inch pounds

M_s = resisting moment of the tensile stresses in the longitudinal steel reinforcement, in inch-pounds

M = bending moment resulting from external forces, in inch-pounds

C = sum of the horizontal compressive stresses in the concrete, in pounds

T = sum of the horizontal tensile stresses in the longitudinal steel reinforcement, in pounds

17-5. Flexure Formulas: Working Stress Design

In accordance with elastic theory, the deformation of any fiber in a beam is directly proportional to its distance from the neutral surface (neutral axis of the cross section), as indicated in Figure 17-2. Likewise, the stresses in the fibers are directly proportional to their deformations as well as to their distances from the neutral axis. By definition,

$$E(\text{modulus of elasticity}) = \frac{\text{unit stress}}{\text{unit deformation}}$$

Hence

$$E_s = \frac{f_s}{\Delta_s} \quad \text{and} \quad E_c = \frac{f_c}{\Delta_c}$$

Transposing,

$$\Delta_s = \frac{f_s}{E_s} \quad \text{and} \quad \Delta_c = \frac{f_c}{E_c}$$

Since the deformations are directly proportional to their distances from the neutral surface,

$$\frac{f_c/E_c}{f_s/E_s} = \frac{kd}{d - kd}$$

or

$$\frac{f_c E_s}{f_s E_c} = \frac{k}{1 - k} \tag{a}$$

Substituting in (a) the value $n = E_s/E_c$,

$$\frac{n f_c}{f_s} = \frac{k}{1 - k} \quad \text{or} \quad \frac{f_c}{f_s} = \frac{k}{n(1 - k)} \tag{b}$$

from which

$$f_c = \frac{f_s k}{n(1 - k)} \tag{17.1}$$

and

$$f_s = \frac{n f_c(1 - k)}{k} \tag{17.2}$$

From (17.2),

$$k = \frac{n f_c(1 - k)}{f_s} \quad \text{or} \quad k = \frac{n - nk}{f_s/f_c}$$

and

$$k = \frac{n}{n + (f_s/f_c)} \qquad (17.3)$$

The compressive stresses in the concrete vary from zero at the neutral surface to f_c at kd distance; hence the average stress is $\frac{1}{2}f_c$. The area of beam in compression is $b \times kd$. Therefore, the sum of all the compressive stresses is

$$C = \frac{1}{2}f_c kbd$$

The forces C and T (Figure 17-2) constitute a mechanical couple, the lever arm of which is jd. Hence for the resisting moment in the concrete,

$$M_c = (\tfrac{1}{2}f_c kbd)\,jd = \tfrac{1}{2}f_c jkbd^2 \qquad (c)$$

Next, consider the resisting moment with respect to the tensile stresses in the steel reinforcement. The area of the steel is A_s, the unit stress is f_s, and therefore T, the sum of all the tensile stresses, is $A_s f_s$. Again the moment arm of the couple is jd; therefore, for the resisting moment of the steel,

$$M_s = A_s f_s jd$$

But

$$p = \frac{A_s}{bd} \quad \text{or} \quad A_s = pbd$$

Therefore,

$$M_s = A_s f_s jd \quad \text{or} \quad M_s = pf_s jbd^2 \qquad (d)$$

The two resisting moments M_c and M_s are, of course, equal in magnitude; hence equating (c) and (d),

$$pf_s jbd^2 = \tfrac{1}{2}f_c jkbd^2$$

from which

$$2pf_s = f_c k \quad \text{or} \quad \frac{f_c}{f_s} = \frac{2p}{k} \qquad (e)$$

Placing the values of f_c/f_s in (b) and (e) equal to each other,

$$\frac{k}{n(1 - k)} = \frac{2p}{k}$$

$$k^2 = 2pn(1 - k) \quad \text{or} \quad k^2 + 2pnk = 2pn$$

Completing the square and solving for k,

$$k = \sqrt{2pn + (pn)^2} - pn \tag{17-4}$$

The resultant of the compressive stresses is $\frac{1}{3}kd$ from the uppermost fiber, and, since the distance between the resultant of the compressive stresses and the resultant of the tensile stresses is jd (Figure 17-2),

$$jd = d - \frac{kd}{3} \quad \text{or} \quad j = 1 - \frac{k}{3} \tag{17-5}$$

Since the magnitudes of the bending moment and the resisting moment must be equal, (c) may be written

$$M = \tfrac{1}{2}f_c jkbd^2 \quad \text{or} \quad d^2 = \frac{M}{\tfrac{1}{2}f_c jkb}$$

and

$$d = \sqrt{\frac{M}{\tfrac{1}{2}f_c jkb}}$$

As a matter of convenience, let $R = \tfrac{1}{2}f_c jk$; then

$$d = \sqrt{\frac{M}{Rb}} \tag{17-6}$$

The bending moment may also be substituted for the resisting moment M_s in (d). Then

$$M = A_s f_s jd$$

and

$$A_s = \frac{M}{f_s jd} \tag{17-7}$$

By definition,

$$p = \frac{A_s}{bd}$$

Therefore,

$$A_s = pbd \tag{17-8}$$

Equation (e) gives

$$2pf_s = f_c k$$

Transposing,

$$p = \frac{kf_c}{2f_s} \tag{17-9}$$

The formulas developed apply to rectangular beams in flexure (bending). They may also be used in the design of slabs, since a floor slab may be considered a rectangular beam with a width several times its depth.

17-6. Allowable Stresses and Formula Coefficients

Because the ingredients of concrete may be combined in various proportions, concrete of different strengths may be produced. As noted earlier, the symbol f_c' represents the specified compressive strength; this may be considered as the ultimate compressive strength of the concrete at the age of 28 days. Commonly used grades of concrete are shown by the f_c' values given in Table 17-1. Allowable working values for the extreme fiber stress are established as a percentage of the specified compressive strength from the relationship $f_c = 0.45 \times f_c'$; some values of f_c are also listed in Table 17-1. The allowable tensile stress in the steel reinforcement is usually 20,000 or 24,000 psi [140,165 MPa] depending on the grade of reinforcing bars used.

Unlike homogeneous beams, the position of the neutral axis in reinforced concrete beams depends on the unit tensile and compressive stresses and on n, the ratio of the modulus of elasticity of steel to the modulus of elasticity of concrete, that is, $n = E_s/E_c$. For the reinforcement, $E_s = 29,000,000$ psi [200 GPa] but E_c varies according to the compressive strength of the concrete; it may be considered equal to $57,000 \times \sqrt{f_c'}$ [$15,100\sqrt{f_c'}$]. Table 17-1 gives values of n for the grades of concrete listed.

17-7. Balanced Reinforcement

A useful reference is the so-called balanced section, which occurs when the exact amount of reinforcing used results in the simultaneous limiting stresses in the concrete and steel. The properties that establish this relationship may be expressed as follows:

$$\text{balanced } k = \frac{1}{1 + f_s/nf_c} \qquad (17\text{-}10)$$

$$j = 1 - \frac{k}{3} \qquad (17\text{-}5)$$

$$p = \frac{f_c k}{2f_s} \qquad (17\text{-}9)$$

$$M = Rbd^2 \qquad (17\text{-}11)$$

in which

$$R = \tfrac{1}{2}kjf_c \qquad (17\text{-}12)$$

[derived from Formula (c) in Sect. 17-5]. If the limiting compression stress in the concrete ($f_c = 0.45 f_c'$) and the limiting stress in the steel are entered in Formula 17-10, the balanced section value for k may be found. Then the corresponding values for j, p, and R may be found. The balanced p may be used to determine the maximum amount of tensile reinforcing that may be used in a section without the addition of compressive reinforcing. If less tensile reinforcing is used, the moment will be limited by the steel stress, the maximum stress in the concrete will be below the limit of $0.45f_c'$, the value of k will be slightly lower than the balanced value, and the value of j slightly higher than the balanced value. These relationships are useful in design for the determination of approximate requirements for cross sections.

Table 17-2 gives the balanced properties for various combinations

TABLE 17-2. Balanced Section Properties for Rectangular Concrete Sections with Tension Reinforcing Only

f_s		f'_c		n	k	j	p	R	
ksi	MPa	ksi	MPa					k-in.	kN·m
16	110	2.0	13.79	11.3	0.389	0.870	0.0109	0.152	1045
		2.5	17.24	10.1	0.415	0.862	0.0146	0.201	1382
		3.0	20.68	9.2	0.437	0.854	0.0184	0.252	1733
		4.0	27.58	8.0	0.474	0.842	0.0266	0.359	2468
20	138	2.0	13.79	11.3	0.337	0.888	0.0076	0.135	928
		2.5	17.24	10.1	0.362	0.879	0.0102	0.179	1231
		3.0	20.68	9.2	0.383	0.872	0.0129	0.226	1554
		4.0	27.58	8.0	0.419	0.860	0.0188	0.324	2228
24	165	2.0	13.79	11.3	0.298	0.901	0.0056	0.121	832
		2.5	17.24	10.1	0.321	0.893	0.0075	0.161	1107
		3.0	20.68	9.2	0.341	0.886	0.0096	0.204	1403
		4.0	27.58	8.0	0.375	0.875	0.0141	0.295	2028

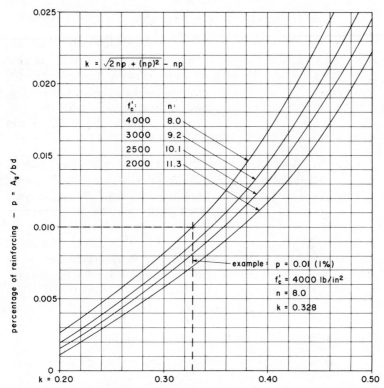

FIGURE 17-3.   k factors for rectangular sections with tension reinforcing only—
as a function of n and p.

of concrete strength and limiting steel stress. The values of n, k, j, and p are all without units. However, R must be expressed in particular units; the unit used in the table is kip-inches (k-in.).

When the area of steel used is less than the balanced p, the true value of k may be determined by the previously derived formula:

$$k = \sqrt{2np - (np)^2} - np \qquad (17\text{-}4)$$

Figure 17-3 may be used to find approximate k values for various combinations of p and n.

17-8. Use of Working Stress Formulas

In the design of concrete beams, two situations commonly occur. The first occurs when the beam is entirely undetermined; that is, the concrete dimensions and the reinforcing are unknown. The second occurs when the concrete dimensions are given, and the required reinforcing for a specific bending moment must be determined. The following examples illustrate the use of the formulas just developed for each of these problems.

Example 1. A rectangular concrete beam of concrete with f'_c of 3000 psi [20.7 MPa] and steel reinforcing with $f_s = 20$ ksi [138 MPa] must sustain a bending moment of 200 k-ft [271 kN-m]. Select the beam dimensions and the reinforcing for a section with tension reinforcing only.

Solution: (1) With tension reinforcing only, the minimum size beam will be a balanced section, since a smaller beam would have to be stressed beyond the capacity of the concrete to develop the required moment. Using Formula (17-11),

$$M = Rbd^2 = 200 \text{ kip-ft [271 kN-m]}$$

Then from Table 17-2, for f'_c of 3000 psi and f_s of 20 ksi,

$$R = 0.226 \text{ (in units of k-in.) [1554 in units of kN-m]}$$

Therefore,

$$M = 200 \times 12 = 0.226(bd^2), \text{ and } bd^2 = 10{,}619$$

$$[M = 271 = 1554(bd^2), \text{ and } bd^2 = 0.1744]$$

(2) Various combinations of b and d may be found; for example,

$$b = 10 \text{ in.}, d = \sqrt{\frac{10{,}619}{10}}$$

$$= 32.6 \text{ in.}, [b = 0.254 \text{ m}, d = 0.829 \text{ m}]$$

$$b = 15 \text{ in.}, d = \sqrt{\frac{10{,}619}{15}}$$

$$= 26.6 \text{ in.}, [b = 0.381 \text{ m}, d = 0.677 \text{ m}]$$

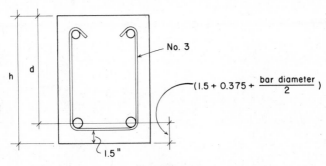

FIGURE 17-4

Although they are not given in this example, there are often some considerations other than flexural behavior alone that influence the choice of specific dimensions for a beam. If the beam is of the ordinary form shown in Figure 17-4, the specified dimension is usually that given as h. Assuming the use of a No. 3 U-stirrup, a cover of 1.5 in. [38 mm], and an average-size reinforcing bar of 1-in. [25-mm] diameter (No. 8 bar), the design dimension d will be less than h by 2.375 in. [60 mm]. Lacking other considerations, we will assume a b of 15 in. [380 mm] and an h of 29 in. [740 mm], with the resulting d of 29 − 2.375 = 26.625 in. [680 mm].

(3) We next use the specific value for d with Formula (17-7) to find the required area of steel A_s. Since our selection is very close to the balanced section, we may use the value of j from Table 17-2. Thus

$$A_s = \frac{M}{f_s jd} = \frac{200 \times 12}{20 \times 0.872 \times 26.625} = 5.17 \text{ in.}^2$$

$$\left[A_s = \frac{271,000}{0.138 \times 0.872 \times 680} = 3312 \text{ mm}^2 \right]$$

Or using the formula for the definition of p and the balanced p value from Table 17-2,

$$A_s = pbd = 0.0129(15 \times 26.625) = 5.15 \text{ in.}^2$$

$$[A_s = 0.0129(380 \times 680) = 3333 \text{ mm}^2]$$

(4) We next select a set of reinforcing bars to obtain this area. (See

TABLE 17-3. Properties of Standard Reinforcing Bars

Size	Nominal Diameter		Nominal Area		Nominal Perimeter		Weight	
	(in.)	(mm)	(in.²)	(mm²)	(in.)	(mm)	(lb/ft)	(kg/m)
3	0.375	9.52	0.11	71	1.178	29.92	0.376	0.560
4	0.500	12.70	0.20	129	1.571	39.90	0.668	0.994
5	0.625	15.88	0.31	200	1.963	49.86	1.043	1.552
6	0.750	19.05	0.44	284	2.356	59.84	1.502	2.235
7	0.875	22.22	0.60	387	2.749	69.82	2.044	3.042
8	1.000	25.40	0.79	510	3.142	79.81	2.670	3.973
9	1.128	28.65	1.00	645	3.544	90.02	3.400	5.060
10	1.270	32.26	1.27	819	3.990	101.35	4.303	6.404
11	1.410	35.81	1.56	1006	4.430	112.52	5.313	7.907
14	1.693	43.00	2.25	1452	5.320	135.13	7.650	11.380
18	2.257	57.33	4.00	2581	7.090	180.09	13.600	20.240

Table 17-3.) As with the beam dimensions, there are other concerns. For the purpose of our example, if we select bars all of a single size the number required will be:

$$\text{For No. 6 bars,} \quad \frac{5.17}{0.44} = 11.75, \quad \text{or 12} \quad \left[\frac{3312}{284} = 11.66\right]$$

$$\text{For No. 7 bars,} \quad \frac{5.17}{0.60} = 8.62, \quad \text{or 9} \quad \left[\frac{3312}{387} = 8.56\right]$$

$$\text{For No. 8 bars,} \quad \frac{5.17}{0.79} = 6.54, \quad \text{or 7} \quad \left[\frac{3312}{510} = 6.49\right]$$

$$\text{For No. 9 bars,} \quad \frac{5.17}{1.00} = 5.17, \quad \text{or 6} \quad \left[\frac{3312}{645} = 5.13\right]$$

$$\text{For No. 10 bars,} \quad \frac{5.17}{1.27} = 4.07, \quad \text{or 5} \quad \left[\frac{3312}{819} = 4.04\right]$$

$$\text{For No. 11 bars,} \quad \frac{5.17}{1.56} = 3.31, \quad \text{or 4} \quad \left[\frac{3312}{1006} = 3.29\right]$$

For all except the No. 11 bars, the requirements for bar spacing would result in the need to place the bars in stacked layers in the 15-in.-wide beam. Although this is possible, it would require some increase in the dimension h to maintain the effective depth of approximately 26.6 in., since the centroid of the steel bar areas would move farther away from the edge of the concrete.

Example 2. A rectangular concrete beam of concrete with f'_c of 3000 psi [20.7 MPa] and steel with f_s of 20 ksi [138 MPa] has dimensions of $b = 15$ in. [380 mm] and $h = 36$ in. [910 mm]. Find the area required for the steel reinforcing for a moment of 200 kip-ft [271 kN-m].

Solution: The first step in this case is to determine the balanced moment capacity of the beam with the given dimensions. If we assume the section to be as shown in Figure 17-4, we may assume an approximate value for d to be h minus 2.5 in. [64 mm], or 33.5 in. [851 mm]. Then with the value for R from Table 17-2,

$$M = Rbd^2 = 0.226 \times 15 \times (33.5)^2 = 3804 \text{ k-in.}$$

or

$$M = \frac{3804}{12} = 317 \text{ k-ft}$$

$$[M = 1554 \times 0.380 \times (0.851)^2 = 428 \text{ kN-m}]$$

Since this value is considerably larger than the required moment, it is thus established that the given section is larger than that required for a balanced stress condition. As a result, the concrete flexural stress will be lower than the limit of $0.45f'_c$, and the section is qualified as being under-reinforced; which is to say that the reinforcing required will be less than that required to produce a balanced section (with moment capacity of 317 k-ft). To find the required area of steel, we use Formula (17-7) just as we did in the preceding example. However, the true value for j in the formula will be something greater than that for the balanced section (0.872 from Table 17-2).

As the amount of reinforcing in the section decreases below the full amount required for a balanced section, the value of k decreases and the value of j increases. However, the range for j is small: from 0.872 up to something less than 1.0. A reasonable procedure is to assume a value for j, find the corresponding required area, and then perform an investigation to verify the assumed value for j, as follows.

Assume $j = 0.90$. Then

$$A_s = \frac{M}{f_s j d} = \frac{200 \times 12}{20 \times 0.90 \times 33.5} = 3.98 \text{ in.}^2$$

and

$$p = \frac{A_s}{bd} = \frac{3.98}{15 \times 33.5} = 0.00792$$

$$\left[A_s = \frac{271,000}{0.138 \times 0.90 \times 850} = 2567 \text{ mm}^2 \right.$$

$$\left. p = \frac{2567}{380 \times 850} = 0.00795 \right]$$

Using this value for p in Figure 17-3, we find $k = 0.313$. Using Formula (17-5), we then determine j to be

$$j = 1 - \frac{k}{3} = 1 - \frac{0.313}{3} = 0.896$$

which is reasonably close to our assumption, so the computed area is adequate for design.

Problem 17-8-A.   A rectangular concrete beam has concrete with $f'_c = 3000$ psi [20.7 MPa] and steel reinforcing with $f_s = 20$ ksi [138 MPa]. Select the beam dimensions and reinforcing for a balanced section if the beam sustains a bending moment of 240 k-ft [325 kN-m].

Problem 17-8-B.   Find the area of steel reinforcing required and select the bars for the beam in Problem 17-8-A if the section dimensions are $b = 16$ in. and $d = 32$ in.

17-9.   The Strength Design Method

Application of the working stress method consists of designing members to *work* in an adequate manner (without exceeding established stress limits) under actual service load conditions. The basic procedure in strength design is to design members to *fail*; thus the ultimate strength of the member at failure (called its design strength) is the only type of resistance considered. Safety in strength design is not provided by limiting stresses, as in the working stress method, but by using a factored design load (called the *required strength*) that is greater than the service load. The code establishes the value of the required strength, called U, as not less than

$$U = 1.4D + 1.7L \tag{17-13}$$

in which D = the effect of dead load
$\qquad\quad L$ = the effect of live load

Other adjustment factors are provided when design conditions involve consideration of the effects of wind, earth pressure, differential settlement, creep, shrinkage, or temperature change.

The design strength of structural members (i.e., their *usable* ultimate strength) is determined by the application of assumptions and requirements given in the code and is further modified by the use of a *strength reduction factor* ϕ as follows:

$\phi = 0.90$ for flexure, axial tension, and combinations of flexure and tension

= 0.75 for columns with spirals
= 0.70 for columns with ties
= 0.85 for shear and torsion
= 0.70 for compressive bearing
= 0.65 for flexure in plain (not reinforced) concrete

Thus although Formula (17-13) may imply a relatively low safety factor, an additional margin of safety is provided by the strength reduction factors.

17-10. Flexure Formulas: Strength Design

Figure 17-5 shows the equivalent rectangular compressive stress distribution in the concrete permitted by the ACI Code for use in the strength design method. As stated in Section 17-2, the rectangular stress block is based on the assumption that, at ultimate load, a concrete stress of $0.85 f_c'$ is uniformly distributed over the compression zone. The dimensions of this zone are the beam width b and the distance a which locates a line parallel to and above the neutral axis. Although we have not yet considered how the value of a is determined, let us turn our attention to Figure 17-5 and develop equations for the theoretical resisting moment M_t.

We observe that the resultant (sum) of the compressive stresses is

$$C = 0.85 f_c' \times b \times a$$

and that it acts at a distance of $a/2$ from the top of the beam. The arm of the resisting moment couple jd then becomes $d - a/2$, and the the-

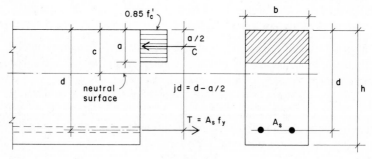

FIGURE 17-5

oretical resisting moment as governed by the concrete is

$$M_t = C\left(d - \frac{a}{2}\right) = (0.85 \, f'_c ba) \times \left(d - \frac{a}{2}\right) \qquad (17\text{-}14)$$

Similarly, the theoretical moment strength as controlled by the steel reinforcement is

$$M_t = T\left(d - \frac{a}{2}\right) = (A_s f_y)\left(d - \frac{a}{2}\right) \qquad (17\text{-}15)$$

If *balanced* conditions exist, that is, if the concrete reaches its full compressive strength when the steel reaches its yield strength, the two equations will be equal to each other, or

$$0.85 \, f'_c ba = A_s f_y = \rho b d f_y \qquad (17\text{-}16)$$

where

$$\rho = \frac{A_s}{bd}$$

Note: The ACI Code uses ρ to indicate steel percent with the strength method, whereas p is used with the working stress method. Then, from Formula (17-16)

$$a = \frac{\rho b d f_y}{0.85 \, f'_c b} = \frac{\rho f_y d}{0.85 \, f'_c} \qquad (17\text{-}17)$$

and

$$\rho = \frac{a}{d} \times \frac{0.85 \, f'_c}{f_y} \qquad (17\text{-}18)$$

The symbol ρ_b is used to denote the balancing ratio of reinforcement and a_b, the depth of the stress block under balanced conditions.

Using this expression for a and considering the strain relationship between concrete and steel, the following formula for the balancing ratio of reinforcement may be derived:

$$\rho_b = \frac{0.85 \, f'_c \beta_1}{f_y} \times \frac{87,000}{87,000 + f_y} \qquad (17\text{-}19)$$

in which β_1 is a coefficient relating the depth of the rectangular stress block to the depth from the compression face to the neutral axis, or $a = \beta_1 \times c$. (See Figure 17-5). The value of β_1 varies with the strength of the concrete. The ACI Code prescribes a value of 0.85 for concrete strengths up to 4000 psi [27.6 MPa] and a reduction of 0.05 for each 1000 psi [6.895 MPa] of strength in excess of 4000 psi, with a minimum value of 0.65. (For example, if $f'_c = 5000$ psi [34.5 MPa], $\beta_1 = 0.80$, and so on.)

By equating Formulas (17-18) and (17-19), we can derive an expression for the balanced value of a/d. Thus

$$\frac{a}{d} \times \frac{0.85\, f'_c}{f_y} = \frac{0.85\, f'_c \beta_1}{f_y} \times \frac{87{,}000}{87{,}000 + f_y}$$

and

$$\frac{a}{d} = \beta_1 \frac{87{,}000}{87{,}000 + f_y} \qquad (17\text{-}20)$$

$$\left[\begin{array}{l} \text{For SI units:} \\[2mm] \dfrac{a}{d} = \beta_1 \dfrac{600{,}000}{600{,}000 + f_y} \\[2mm] \text{when stress is in MPa} \end{array}\right]$$

Referring to Formula (17-14), we may derive another form for this equation as follows:

$$M_t = 0.85\, f'_c ba\left(d - \frac{a}{2}\right)$$

$$= 0.85\, f'_c b \frac{a}{d} d\left(1 - \frac{1}{2}\frac{a}{d}\right)$$

$$= bd^2\left[0.85 f'_c\left\{\frac{a}{d} - \frac{1}{2}\left(\frac{a}{d}\right)^2\right\}\right]$$

$$= Rbd^2 \qquad (17\text{-}21)$$

where

$$R = 0.85 f'_c\left[\frac{a}{d} - \frac{1}{2}\left(\frac{a}{d}\right)^2\right] \qquad (17\text{-}22)$$

TABLE 17-4. Balanced Section Properties for Rectangular Concrete Sections with Tension Reinforcing Only: Strength Design[a]

f'_c		$f_y = 40$ ksi [276 MPa]						$f_y = 60$ ksi [414 MPa]					
psi	MPa	Balanced a/d	Usable a/d (75% Balance)	Usable ρ		Usable R		Balanced a/d	Usable a/d (75% Balance)	Usable ρ		Usable R	
						k-in.	kN-m					k-in.	kN-m
2000	13.79	0.5823	0.4367	0.0186		0.580	4000	0.5031	0.3773	0.0107		0.520	3600
2500	17.24	0.5823	0.4367	0.0232		0.725	5000	0.5031	0.3773	0.0137		0.650	4500
3000	20.69	0.5823	0.4367	0.0278		0.870	6000	0.5031	0.3773	0.0160		0.781	5400
4000	27.58	0.5823	0.4367	0.0371		1.161	8000	0.5031	0.3773	0.0214		1.041	7200
5000	34.48	0.5480	0.4110	0.0437		1.388	9600	0.4735	0.3551	0.0252		1.241	8600

[a]See Section 17-10 for derivation of formulas used to obtain table values.

Formulas (17-19), (17-20), and (17-22) may be used to derive factors for balanced sections which can be used in the design of beams. Table 17-4 contains a compilation of these factors for five values of concrete strength and two values of steel yield strength. The use of this material is demonstrated in the following sections.

17-11. Flexure Notation: Strength Design

This section lists symbols used in the design of rectangular beams for bending when the strength design method is employed. The notation follows generally that contained in the 1983 ACI Code.

a = depth of equivalent rectangular stress block

A_s = area of tension reinforcement (in.^{2})

b = width of compression face of member

c = distance from extreme compression fiber to neutral axis

C = sum of compressive stresses in the concrete

d = effective depth, the distance from extreme compression fiber to centroid of tension reinforcement

D = dead loads or their related internal moments and forces

f'_c = specified compressive strength of concrete (psi)

f_s = calculated stress in reinforcement at service loads

f_y = specified yield strength of reinforcement (psi)

h = total depth or thickness of member

L = live loads or their related internal moments and forces

M_t = theoretical moment strength of a section (in-lb)

M_u = applied design moment at a section

T = sum of tensile stresses in the reinforcement

U = required strength to resist design loads or their related internal moments and forces

β_1 = a coefficient relating the depth of the rectangular stress block to the depth from the compression face to the neutral axis, $a = \beta_1 \times c$

ρ = ratio of tension reinforcement = A_s/bd

ρ_b = reinforcement ratio producing balanced conditions

ρ_{max} = maximum reinforcement ratio = $0.75\rho_b$

ρ_{min} = minimum reinforcement ratio = $200/f_y$

ϕ = capacity reduction factor

ω = $\rho f_y/f'_c$

17-12. Use of Strength Design Formulas

Use of the strength design formulas derived in Section 17-10 is illustrated in the following examples.

Example 1.   The service load bending moments on a rectangular beam 10 in. [254 mm] wide are 58 k-ft [78.6 kN-m] for dead load and 38 k-ft [51.5 kN-m] for live load. If $f'_c = 4000$ psi [27.6 MPa] and $f_y = 60$ ksi [414 MPa], determine the depth of the beam and the required area of tension reinforcing.

Solution: (1) The required ultimate moment strength M_u is first determined in accordance with Section 17-9.

$$U = 1.4D + 1.7L$$

$$M_u = 1.4(M_{DL}) + 1.7(M_{LL})$$

$$= 1.4(58) + 1.7(38) = 146 \text{ k-ft}$$

$$[M_u = 1.4(78.6) + 1.7(51.5) = 198 \text{ kN-m}]$$

(2) To find the required design moment strength M_t we apply the capacity reduction factor $\phi = 0.90$ (Section 17-9) and the relationship $M_u = \phi M_t$; thus

$$M_t = \frac{M_u}{\phi} = \frac{146}{0.90} = 162 \text{ k-ft} \text{ or } 1944 \text{ k-in.}$$

$$\left[M_t = \frac{198}{0.90} = 220 \text{ kN-m} \right]$$

(3) The maximum usable reinforcement ratio as given in Table 17-4 is $\rho = 0.0214$. If a balanced section is used, we may thus determine the required area of reinforcement from the relationship

$$A_s = \rho b d$$

Although there is nothing especially desirable about a balanced section, it does represent the beam section with least depth if tension reinforcing only is used. We will therefore proceed to find the required balanced section for this example.

(4) To determine the required effective depth d, we use Formula (17-21) from Section 17-10; thus

$$M_t = Rbd^2$$

With the value of $R = 1.041$ from Table 17-4,

$$M_t = 1944 = 1.041(10)(d)^2$$

and

$$d = \sqrt{\frac{1944}{1.041(10)}} = \sqrt{186.7} = 13.66 \text{ in.}$$

$$\left[d = \sqrt{\frac{220}{7200(0.254)}} = 0.347 \text{ m} \right]$$

(5) If this value is used for d, the required steel area may be found as

$$A_s = \rho bd = 0.0214 \times 10 \times 13.66 = 2.92 \text{ in.}^2$$

The ACI Code requires a minimum ratio of reinforcing as follows:

$$\rho_{min} = \frac{200}{f_y} = \frac{200}{60,000} = 0.0033$$

which is clearly not critical for this example.

Selection of the actual beam dimensions and the actual number and size of reinforcing bars would involve various considerations, as discussed in Section 17-8 and illustrated in Fig. 17-4.

If there are reasons, as there often are, for not selecting the least deep section with the greatest amount of reinforcement, a slightly different procedure must be used, as illustrated in the following example.

Example 2. Using the same data as in Example 1, find the reinforcing required if the desired beam section has $b = 10$ in. [254 mm] and $d = 18$ in. [457 mm].

Solution: The first two steps in this situation would be the same as in Example 1—to determine M_u and M_t. The next step would be to determine whether the given section is larger than, smaller than, or equal to a balanced section. Since this investigation has already been done in Example 1, we may observe that the 10×18-in. section is larger than a balanced section. Thus the actual value of a/d will be less than the balanced section value of 0.3773. The next step would then be as follows:

(4) Estimate a value for a/d—something smaller than the balanced

value of 0.3773. For example, try $a/d = 0.25$. Then

$$a = 0.25d = 0.25(18) = 4.5 \text{ in. } [114 \text{ mm}]$$

With this assumed value for a, we may use Formula (17-15) to find a value for A_s.

(5) Referring to Section 17-10 and Figure 17-5,

$$A_s = \frac{M_t}{f_y(d - a/2)} = \frac{1944}{60(15.75)} = 2.057 \text{ in.}^2$$

(6) We next test to see if the estimate for a/d was close by finding a/d using Formula (17-18) of Section 17-10. Thus:

$$\rho = \frac{A_s}{bd} = \frac{2.057}{10 \times 18} = 0.0114$$

and

$$\frac{a}{d} = \rho \frac{f_y}{0.85f'_c} = 0.0114 \frac{60}{0.85(4)} = 0.202$$

$$a = 0.202(18) = 3.63 \text{ in.}, \quad d - \frac{a}{2} = 16.2 \text{ in.}$$

If we replace the value for $d = a/2$ that was used earlier with this new value, the required value of A_s will be slightly reduced. In this example, the correction will be only a few percent. If the first guess for a/d had been way off, it may justify a second run through steps 4, 5, and 6 to get closer to an exact answer.

Problem 17-12-A.   A rectangular beam has concrete with $f'_c = 3000$ psi [20.7 MPa] and steel reinforcing with $f_y = 40$ ksi [276 MPa]. Using strength methods, find the depth required and the area of steel required for a balanced section if the beam width is 16 in. [406 mm]. Service load dead load moment is 140 k-ft [190 kN-m], and live load moment is 100 k-ft [136 kN-m].

Problem 17-12-B.   Find the steel area required for the section in Problem 17-12-A if the section dimensions are $b = 16$ in. [406 mm] and $d = 32$ in. [812 mm].

(Compare the answers obtained in these problems with those obtained in Problems 17-8-A and 17-8-B, in which the problem data is similar but the work is done by the working stress method.)

17-13. Shear in Beams

Let us consider the case of a simple beam with uniformly distributed load and end supports that provide only vertical resistance (no moment restraint). The distribution of internal shear and bending moment are as shown in Figure 8-4. For flexural resistance, it is necessary to provide longitudinal reinforcing bars near the bottom of the beam. These bars are oriented for primary effectiveness in resistance to tension stresses that develop on a vertical (90°) plane (which is the case at the center of the span, where the bending moment is maximum and the shear approaches zero).

Under the combined effects of shear and bending, the beam tends to develop tension cracks as shown in Figure 17-6a. Near the center of the span, where the bending is predominant and the shear approaches zero, these cracks approach 90°. Near the support, however, where the shear predominates and bending approaches zero, the critical tension stress plane approaches 45°, and the horizontal bars are only partly effective in resisting the cracking.

For beams, the most common form of shear reinforcement consists of a series of U-shaped bent bars (Figure 17-6c), placed vertically and spaced along the beam span, as shown in Figure 17-6b. These bars are

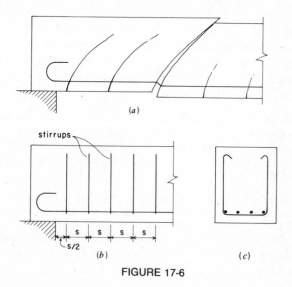

FIGURE 17-6

intended to provide a vertical component of resistance, working in conjunction with the horizontal resistance provided by the flexural reinforcement. To develop tension near the support face, the horizontal bars must be bonded to the concrete beyond the point where the stress is developed. Where the beam ends extend only a short distance over the support (a common situation), it is often necessary to bend or hook the bars, as shown in Figure 17-6.

17-14. Beam Shear: General Design Considerations

The following are some of the general considerations and code requirements that apply to current practices of design for beam shear.

Concrete Capacity.   Whereas the tensile strength of the concrete is ignored in design for flexure, the concrete is assumed to take some portion of the shear in beams. If the capacity of the concrete is not exceeded—as is sometimes the case for lightly loaded beams—there may be no need for reinforcing. The typical case, however, is as shown in Figure 17-7, where the maximum shear V exceeds the capacity of the concrete alone V_c, and the steel reinforcing is required to absorb the excess, indicated as the shaded portion in the shear diagram.

Minimum Shear Reinforcing.   Even when the maximum computed shear stress falls below the capacity of the concrete, the present code requires the use of some minimum amount of shear reinforcing. Exceptions are made in some situations, such as for slabs and very shallow beams. The objective is essentially to toughen the structure with a small investment in additional reinforcing.

Type of Stirrup.   The most common stirrups are the simple U shape or closed forms shown in Figure 17-4, placed in a vertical position at intervals along the beam. It is also possible to place stirrups at an incline (usually 45°), which makes them somewhat more effective in

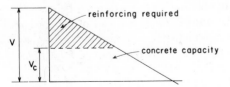

FIGURE 17-7

direct resistance to the potential shear cracking near the beam ends. In large beams with excessively high unit shear stress, both vertical and inclined stirrups are sometimes used at the location of the greatest shear.

Size of Stirrups.   For beams of moderate size, the most common size for U-stirrups is a No. 3 bar. These bars can be bent relatively tightly at the corners (small radius of bend) to fit within the beam section. For larger beams, a No. 4 bar is sometimes used, its strength (as a function of its cross-sectional area) being almost twice that of a No. 3 bar.

Spacing of Stirrups.   Stirrup spacings are computed (as discussed in the following sections) on the basis of the amount of reinforcing required for the unit shear stress at the location of the stirrups. A maximum spacing of $d/2$ (i.e., one-half the effective beam depth d) is specified in order to assure that at least one stirrup occurs at the location of any potential diagonal crack. When shear stress is excessive, the maximum spacing is limited to $d/4$.

Critical Maximum Design Shear.   Although the actual maximum shear value occurs at the end of the beam, the code permits the use of the shear stress at a distance of d (effective beam depth) from the beam end as the critical maximum for stirrup design. Thus, as shown in Figure 17-8, the shear requiring reinforcing is slightly different from that shown in Figure 17-7.

Total Length for Shear Reinforcing.   On the basis of computed shear stresses, reinforcing must be provided along the beam length for the distance defined by the shaded portion of the shear stress diagram shown in Figure 17-8. For the center portion of the span, the concrete is theoretically capable of the necessary shear resistance without the assistance of reinforcing. However, the code requires that some reinforcing be provided for a distance beyond this computed cutoff point. The 1963 ACI Code required that stirrups be provided for a distance equal to the effective depth of the beam beyond the cutoff point. The 1983 ACI Code requires that minimum shear reinforcing be provided as long as the computed shear stress exceeds one-half of the capacity of the concrete. However it is established, the total extended range over which reinforcing must be provided is indicated as R on Figure 17-8.

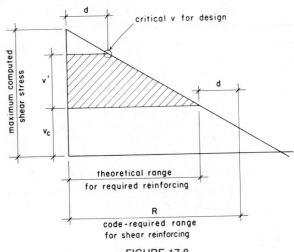

FIGURE 17-8

17-15.   Design of Shear Reinforcement: Strength Design

The requirements and procedures for strength design are essentially similar to those for working stress design. The principal difference is in the use of ultimate resistance as opposed to working stresses at service loads. The basic requirement in strength design is that the modified ultimate resistance of the section be equal to or greater than the factored load. This condition is stated as

$$V_u \leq \phi V_n$$

in which

V_u = the factored shear force at the section
V_n = the nominal shear strength of the section

The nominal strength is defined as

$$V_n = V_c + V_s$$

in which

V_c = nominal strength provided by concrete
V_s = nominal strength provided by reinforcing

The term *nominal strength* is used to differentiate between the computed resistances and the usable value of total resistance, which is reduced for design by the strength reduction factor ϕ. (See Section 17-9 for a general discussion of strength design.)

For members subjected to shear and flexure only, the nominal concrete strength is defined as

$$V_c = 2 \sqrt{f'_c}\, bd$$

Translated into unit stress terms, this means that the limiting nominal shear stress in the concrete is $2 \sqrt{f'_c}$, and when reduced by ϕ, the limiting *working* ultimate strength is $0.85 \times 2 \sqrt{f'_c} = 1.7 \sqrt{f'_c}$.

When shear reinforcing consists of vertical stirrups, the nominal reinforcing strength is defined as

$$V_s = \frac{A_v f_y d}{s}$$

with a limiting value for V_s established as

$$V_s = 8 \sqrt{f'_c}\, bd$$

The following example illustrates the use of strength design methods for shear reinforcing.

Example.   Using strength design methods, determine the spacing required for No. 3 U-stirrups for the beam shown in Figure 17-9. Use $f'_c = 3$ ksi [20.7 MPa] and $f_y = 50$ ksi [345 MPa].
Solution: The loads shown in Figure 17-9a are service loads. These must be converted to *factored loads* for strength design, as discussed in Section 17-9. We thus determine the factored load to be

$$W_u = 1.4(\text{dead load}) + 1.7(\text{live load})$$

$$= 1.4(40) + 1.7(40)$$

$$= 124 \text{ k}$$

The maximum shear force is thus 62 k, and the shear diagram for one-half the beam is as shown in Figure 17-9c. The critical value for V_u at 24 in. (effective beam depth) from the support is determined from proportionate triangles to be 46.5 k. The usable capacity of the con-

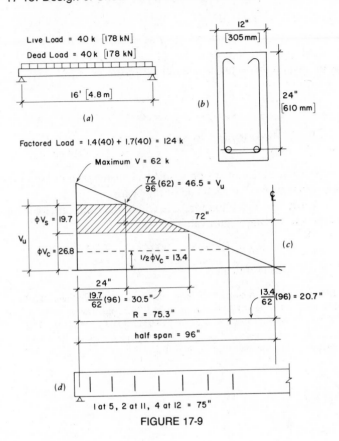

FIGURE 17-9

crete is determined as

$$\phi V_c = \phi 2 \sqrt{f_c'} \, bd$$

$$= 0.85(2 \sqrt{3000})(12 \times 24)$$

$$= 26,816 \text{ lb or approximately } 26.8 \text{ k}$$

and for the reinforcing

$$\phi V_s = 46.5 - 26.8 = 19.7 \text{ k}$$

Therefore,

$$V_s = \frac{19.7}{\phi} = \frac{19.7}{0.85} = 23.18 \text{ k}$$

and the required spacing is determined from

$$V_s = \frac{A_v f_y d}{s}$$

$$s = \frac{A_v f_y d}{V_s} = \frac{0.22 \times 50 \times 24}{23.18} = 11.4 \text{ in.}$$

A possible choice of stirrup spacings is that shown in Figure 17-9d, using seven stirrups at each end of the beam.

To verify that the value for V_s is within the limit previously given, we compute the maximum value of

$$V_s = 8 \sqrt{f_c'} \, bd = 8 \sqrt{3000} \times 12 \times 24 = 126 \text{ k}$$

which is far from critical.

Problem 17-15-A. A concrete beam similar to that shown in Figure 17-9 sustains a total load of 60 kips [267 kN] on a span of 24 ft [7.32 m]. Determine the layout for a set of No. 3 U-stirrups using $f_s = 20$ ksi [138 MPa] and $f_c' = 3000$ psi [20.7 MPa]. The section dimensions are $b = 12$ in. [305 mm] and $d = 26$ in. [660 mm]. (Assume the total load to be one-half LL and one-half DL.)

18

Retaining Walls and Dams

IIIIIIIIIIIIIIIIIIIIIIIIIIIIIIIIIIIIIIIIIIIIII

18-1. General Considerations

A *retaining wall* is a wall employed to resist the lateral pressure (thrust) of earth or other granular material. A *dam* is a retaining wall used to resist the lateral pressure of water or other liquid.

In general, there are three types of retaining walls. The *gravity wall*, illustrated in Figure 18-1a and b, is made of such proportions that its weight alone is sufficient to resist the thrust of the retained material. The *cantilever wall*, shown in Figure 18-1c, d, and e, is constructed of reinforced concrete. The types shown in Figure 18-1c and e make use of the weight of the earth to prevent the wall from overturning at its outer edge. Figure 18-1f shows the *counterfort wall*. It is similar to the *cantilever wall* with the exception that vertical triangular-shaped cross walls are tied to the base and vertical slab at regular intervals. The shapes in which retaining walls are made depend on conditions. Sometimes the exposed surface of the wall shown in Figure 18-1a is a sloping surface. The base slab in the cantilever and counterfort walls is sometimes extended beyond the exposed face, as shown in Figure 18-1e and f; this helps to prevent overturning.

The type of wall selected for a specific situation depends on many factors. Low walls are invariably of the gravity type and are constructed of brick, stone masonry, or unreinforced concrete. The cantilever wall is used for walls of intermediate height up to about 20 ft, with the counterfort wall frequently employed for greater heights.

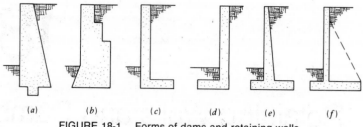

(a)   (b)   (c)   (d)   (e)   (f)

FIGURE 18-1. Forms of dams and retaining walls.

18-2. Earth Pressure

When investigating the stability of a retaining wall, it is necessary to determine the resultant of the forces corresponding to the weight of the wall and the force that results from the retained earth, the thrust. When the dimensions of the wall are known, its weight may be determined accurately. The earth pressure, however, depends on several factors, the types of retained material, sand, broken stone, gravel, clay, etc. When sand or loose earth is deposited on a flat surface, it does not spread out as a liquid but piles up in a mound. This piling up is caused by friction between the individual particles as they slide one on the other. The slope of the side of such a mound of material is called the *slope of repose*, and the angle between the surface of the material and the horizontal is known as the *angle of repose*. The material within the angle of repose exerts no pressure on a retaining wall. The angle of repose varies with different materials, but for average conditions retained soil is assumed to have a slope of 1.5 to 1, which corresponds to an angle of 33° 41′.

When the surface of the retained earth is horizontal, as shown in Figure 18-2a, P, the resultant earth pressure, is assumed to be horizontal at a height of $h/3$ from the base of the retained earth. Its magnitude is

$$P = 0.286 \frac{wh^2}{2}$$

in which

$P =$ the magnitude of the resultant earth pressure in pounds
$w =$ the weight of the retained soil or other material in pounds per cubic foot
$h =$ the height of the retained earth in feet

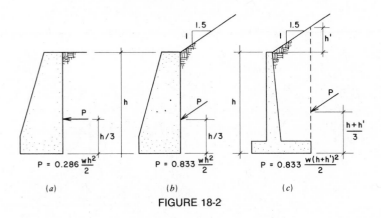

FIGURE 18-2

When the wall is required to retain a *surcharge*, a slope of earth above the top of the wall (Figure 18-2*b* and *c*), the resultant earth pressure is

$$P = 0.833 \; \frac{wh^2}{2}$$

For the three retaining walls in Figure 18-2, the thrust of the earth P is shown in direction, line of action, and magnitude. Note that the direction of the thrust is parallel to the upper surface of the retained earth.

In designing a retaining wall, it is customary to consider a strip of wall 12 in. in length. As an example, if a concrete retaining wall has a cross-sectional area of 3×9 ft, the cross section contains 27 sq ft. hence a strip of wall 12 in. in length contains 27 cu ft, and its weight is 27×150, or 4050 lb.

18-3.   Resultant of Earth Pressure and Wall Weight

The resultant of a system of forces is a single force that produces the same effect as the system of individual forces. If the system is composed of two nonparallel forces, the resultant passes through the point of intersection of their lines of action. The magnitude, direction, and line of action of the resultant can be found graphically by constructing the parallelogram of forces as explained in Section 2-7.

Figure 18-3*a* represents a cross section of a retaining wall whose weight is W. The weight W is downward and acts through the centroid

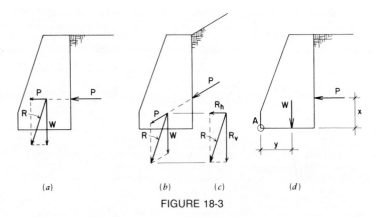

(a)   (b)   (c)   (d)

FIGURE 18-3

of the cross-sectional area. P is the pressure of the earth on the wall. The line of action of P is extended until it meets the line of action of W. At this point a parallelogram of forces of P and W is constructed, and R, the resultant, is determined in magnitude, direction, and line of action. For this particular case note that the vertical component of R is W and that the horizontal component of R is P.

Figure 18-3b illustrates a wall retaining soil on which there is a surcharge. Hence P, the thrust of the earth, is not horizontal. By constructing a parallelogram of forces, R, the resultant of P and W is determined. R is again drawn in Figure 18-3c. The horizontal and vertical components of R are R_H and R_V, respectively. They are not P and W as in Figure 18-3a. When the two components of the resultant are determined, we can proceed to investigate the stability of the wall.

18-4. Stability of Retaining Walls

In general, a retaining wall may fail in one of three ways: (1) by overturning, (2) by undue settlement at the toe, and (3) by sliding horizontally on its base.

The tendency of the wall to overturn is measured by the moment of the force P about point A (Figure 18-3d); the *overturning moment* is $P \times x$. The force W multiplied by its lever arm y is $W \times y$ and is called the *stabilizing moment*. The stabilizing moment divided by the overturning moment determines the factor of safety against overturning; this ratio should be equal at least to 2.

The point at which the resultant of W and P cuts the base of the retaining wall determines the pressure on the foundation bed at the toe of the wall (point A in Figure 18-3d). If the resultant cuts the base of the wall at the center of its width, the pressure on the foundation wall is uniformly distributed. If it cuts the base at any other point, there is an unequal distribution of pressure, and the maximum pressure may be found, as explained in Section 20-1. For important work, test borings should be made to determine the allowable pressure on the foundation bed. Table 18-1 is given as a reference; the allowable bearing capacities in this table are taken from various building codes. Regardless of the magnitude of the maximum pressure, it is good practice to have the wall of such proportions that the resultant falls within the middle third of the base.

The force that tends to cause the retaining wall to slide horizontally is the horizontal component of the resultant. The force resisting this tendency to slide is the product of the vertical component of the resultant and the coefficient of friction of the material composing the foundation bed. Average coefficients of friction of masonry on various foundation beds are: on wet clay, 0.3; on dry clay, 0.5; on sand, 0.4;

TABLE 18-1. Allowable Bearing Capacities of Foundation Beds (ksf or kPa)

Foundation Material	Allowable Vertical Bearing Pressure	
	ksf	kPa
Dry sand, loose	1–2	50–100
Firm	2–4	100–200
Dense	4–8	200–400
Saturated sand, loose	0.5–1	25–50
Firm	1–3	50–150
Dense	3–6	150–300
Clay, soft	0–1.5	0–75
Firm	1.5–2.5	75–125
Hard	4–8	200–400
Gravel, loose	4–6	200–300
Dense	6–12	300–600
Hardpan or hard shale	10–20	500–1,000
Rock, layered, fractured	10–30	500–1,500
Massive, some seams	30–80	1500–4,000
Sound, hard	80–200	4000–10,000

on gravel, 0.6. The factor of safety against sliding is the resisting force divided by the force that tends to cause sliding. A minimum value for the factor of safety is considered to be 1.5. When this value is exceeded, the width of the wall may be increased, thus increasing its weight, or a key may be formed on the base as shown in Figure 18-1a.

18-5.   Stability of a Dam

A dam was defined earlier as a retaining wall used to resist the lateral pressure of water or other liquid. Liquid pressure is always perpendicular to the retaining surface. If the face of a dam is vertical, the water pressure is horizontal, and, with the exception of magnitude, we have a condition similar to that shown in Figure 18-2a.

The intensity of water pressure varies directly with the depth, from zero at the surface to a maximum at the bottom of the water column,

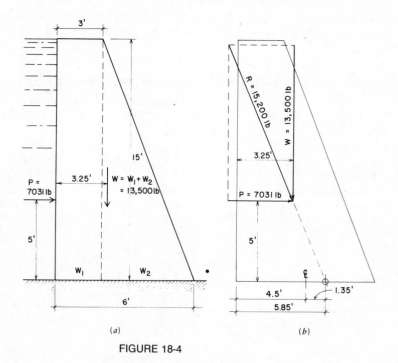

(a)   (b)

FIGURE 18-4

and the resultant of the total horizontal pressure on the face of the dam acts at one-third the distance from the base of the dam to the top.

Example.   Investigate the stability of a masonry dam having the cross section shown in Figure 18-4a. The section is 15 ft high, 3 ft wide at the top, and 9 ft wide at the base. The dam has dense gravel for a foundation bed and retains a maximum depth of water equal to 15 ft. *Solution:* (1) We work with a strip of dam 1 ft long. Recalling that water weighs 62.5 lb/ft^3, the pressure at the base of the dam (15 ft below the upper surface of the water) is $15 \times 62.5 = 937.5$ psf. Since the pressure at the top of the dam is zero, the *average* pressure is half this value, and the total horizontal water pressure on the 1-ft length of dam is

$$\frac{15 \times 62.5}{2} \times 15 = 7031 \text{ lb}$$

This is force P; it acts a distance of $15 \div 3 = 5$ ft above the base of the dam, as shown in Figure 18-4a.

(2) To find the weight of the 1-ft length of dam, we divide the cross section into two parts, W_1 and W_2, the rectangle and triangle as shown. Assuming that the dam is of concrete weighing 150 lb/ft^3, the weight $W_1 = 3 \times 15 \times 1 \times 150$, or 6750 lb. Its moment with respect to the left face of the wall is 6750×1.5, or 10,125 ft-lb. Similarly, the weight for area W_2 is

$$\frac{6 \times 15}{2} \times 1 \times 150 = 6750 \text{ lb}$$

Its moment arm about an axis at the left side of the wall is 6750×5, or 33,750 ft-lb. These computations are recorded in Table 18-2. The total weight of the 1-ft strip of dam is 13,500 lb.

TABLE 18-2. Computations Summary

Section	Weight (lb)	Moment Arm	Moment (lb)
W_1	$3 \times 15 \times 1 \times 150 = 6{,}750$	1.5′	$6750 \times 1.5 = 10{,}125$
W_2	$\dfrac{6 \times 15}{2} \times 1 \times 150 = 6{,}750$	5′	$6750 \times 5 \quad = 33{,}750$
	Total weight, $W = 13{,}500$		Sum of moments $= 43{,}875$

(3) To find the line of action of this force (the position of the centroid of the cross section), we can employ the principle, "the sum of the moments of the parts is equal to the moment of the whole." Let us call x feet the distance of the centroid from the left side of the section. Then

$$13,500 \times x = 43,875 \quad \text{and} \quad x = 3.25 \text{ ft (See Fig. 18-4}a)$$

Now that P and W have been established, the line of action of P is extended until it meets the line of action of W, Figure 18-4b. At this point we construct the parallelogram of forces. By scaling, the resultant R is found to be 15,200 lb. Its horizontal and vertical components are 7031 and 13,500 lb, respectively. We also find by scaling, that the line of action of the resultant cuts the base of the dam at 5.85 ft from the left side. Since the base is 9 ft in width, the resultant cuts the base at 1.35 ft from the center line. This is within the middle third.

(4) The overturning moment about the toe, the lower right-hand edge of the dam, is 7031×5, or 35,155 ft-lb. The moment resisting this tendency to overturn is $13,500 \times 5.75$, or 77,625 ft-lb. The factor of safety against overturning is $77,625 \div 35,155$, or 2.2, and is acceptable.

(5) Since the foundation bed is of gravel, we know the allowable bearing capacity to be 6 to 12 ksf (Table 18-1). Now let us find the maximum pressure on the foundation bed. This pressure will occur at the right-hand side of the base, and the magnitude is found (see Section 20-1) by the formula

$$f_1 = \frac{P}{A}\left(1 + \frac{6e}{d}\right)$$

Then

$$f_1 = \frac{13,500}{9}\left(1 + \frac{6 \times 1.35}{9}\right) \quad \text{and} \quad f_1 = 2850 \text{ psf}$$

As the allowable pressure is 6 to 12 ksf, there will be no undue settlement.

(6) The force tending to cause sliding horizontally is 7031 lb. The force resisting sliding is 13,500 lb times the coefficient of friction for the gravel foundation bed (0.6), or $13,500 \times 0.6 = 8100$ lb. This gives a factor of safety against sliding of $8100 \div 7031 = 1.15$, which

is less than the generally accepted minimum value of 1.5. To increase sliding resistance, a key might be built on the base as shown in Figure 18-1a, or the thickness of the dam might be increased.

18-6. Stability of a Retaining Wall with Surcharge

The water pressure on the dam section investigated in the preceding Section was a horizontal force. When a retaining wall is loaded with a surcharge the thrust is oblique with the horizontal, as shown in Figure 18-2b and c. Figure 18-2c represents the profile of a cantilever reinforced concrete retaining wall. In the design of such a wall, proper proportions of the base and vertical slab are first determined to ensure stability. Then these structural elements are further investigated with respect to reinforcement required and critical stresses in the concrete. Our concern here is limited to the stability of a trial section for a proposed cantilever retaining wall.

Example. A retaining wall has a total height of 12 ft and retains not only the fill in back of the wall but also a surcharge, the slope of which is 1.5 to 1.0. The wall will be of the cantilever type (Figure 18-2c) constructed of reinforced concrete, and the foundation bed is dense dry sand. The configuration and dimensions of a trial cross section are shown in Figure 18-5a. Investigate this trial section for stability, taking the weight of concrete as 150 lb/ft^3 and that of the retained earth as 100 lb/ft^3.

Solution: (1) The resultant earth pressure is parallel to the slope of the surcharge (Figure 18.2c), and its magnitude is

$$P = 0.833 \frac{w(h + h')^2}{2} = 0.833 \times \frac{100(12 + 3)^2}{2} = 9370 \text{ lb}$$

Its line of action passes through a point at $(12 + 3)/3 = 5$ ft above the inside base of the wall. See Figure 18-5a.

(2) To find the combined weight of the wall and the earth above the base, the section is divided into the six components, rectangles and triangles, shown in Figure 18-5a. To establish the line of action of this vertical force, we take the moments of the weights of the various parts with respect to an axis passing through the left side of the base, line AB. Thus the weight of the base, $ABCD$ is $(1.5 \times 8 \times 1 \times 150)$, or

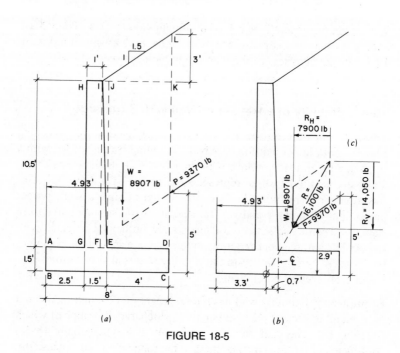

FIGURE 18-5

1800 lb. Its centroid from line AB is $8 \div 2$, or 4 ft, and its moment is 1800×4, or 7200 ft-lb.

In a similar manner, the weights and moments of the various parts are computed and recorded in Table 18-3. (Bear in mind that the centroid of a triangle lies at a point one-third the distance from its base to the apex.) Adding the weights of the various parts, we find that the total weight is 8907 lb. The sum of the moments of the parts is equal to the moment of the whole, and, if we let x feet be the distance from the center of gravity of the resultant force of 8907 lb to the left side of the base,

$$8907 \times x = 43{,}969 \quad \text{and} \quad x = 4.93 \text{ ft}$$

Thus we have established the magnitude and lines of action of the total weight, $W = 8907$ lb, and the thrust of the retained earth, $P = 9370$ lb, as shown in Figure 18-5a.

(3) To avoid confusion of lines the wall is redrawn in Figure 18-5b with forces W and P shown in their proper positions. The resultant of

TABLE 18-3. Computations Summary

Section	Weight (lb)	Moment Arm	Moment (lb)
ABCD	$1.5 \times 8 \times 1 \times 150 = 1800$	$8 \times \frac{1}{2} = 4'$	$1800 \times 4 = 7,200$
GFIH	$1 \times 10.5 \times 1 \times 150 = 1575$	$2.5 + 0.5 = 3'$	$1575 \times 3 = 4,725$
FEI	$0.5 \times 10.5 \times \frac{1}{2} \times 1 \times 150 = 394$	$3.5 + 0.17 = 3.7'$	$394 \times 3.7 = 1,458$
EJI	$0.5 \times 10.5 \times \frac{1}{2} \times 1 \times 100 = 263$	$3.5 + 0.3 = 3.8'$	$263 \times 3.8 = 999$
EDKJ	$4 \times 10.5 \times 1 \times 100 = 4200$	$4 + 2 = 6'$	$4200 \times 6 = 25,200$
KLJ	$4.5 \times 3 \times \frac{1}{2} \times 1 \times 100 = 675$	$3.5 + 3 = 6.5'$	$675 \times 6.5 = 4,387$
	Total weight $= 8907$		Sum of moments $= 43,969$

these two forces acts through their point of intersection. Therefore P is extended on its line of action until it meets the line of action of W. At this point we construct the parallelogram of forces, thus establishing the resultant R. Redrawing R at (c) in the figure, the vertical and horizontal components are found by scaling to be 14,050 and 7900 lb, respectively.

(4) The force that tends to overturn the wall about point B (Figure 18-5a) is 7900 lb, and its moment arm is 2.9 ft. Hence $7900 \times 2.9 = 22,910$ ft-lb, the overturning moment. The force resisting the tendency to overturn is 14,050 lb, and its moment arm is 4.93 ft. Thus $14,050 \times 4.93 = 69,266$ ft-lb. Then the factor of safety against overturning is $69,266 \div 22,910$, or 3.02. This is acceptable since it exceeds 2.

(5) In Figure 18-5b we see that the resultant of W and P intersects the base of the wall 3.3 ft from the left side. Since the base is 8 ft wide, the distance of this point from the center line is $(8 \div 2) - 3.3 = 0.7$ ft. This indicates that the resultant cuts the base within the middle third and the formula (Section 20-1) used to determine the maximum pressure on the foundation bed, the pressure at point B, is

$$f_1 = \frac{P}{A}\left(1 + \frac{6e}{d}\right)$$

Then

$$f_1 = \frac{14,050}{8}\left(1 + \frac{6 \times 0.7}{8}\right) = 2680 \text{ psf}$$

The foundation bed is dense dry sand, and from Table 18-1 we find that the allowable bearing capacity of the soil is 4 to 6 ksf. Since the maximum pressure is only 2680 psf, it is well within the allowable.

(6) The force that tends to cause the wall to slide is the horizontal component of the resultant of the forces acting on the wall. This horizontal component is 7900 lb (Figure 18-5c). The force resisting the tendency to slide is the vertical component of the resultant, 14,050 lb, multiplied by the coefficient of friction of gravel, 0.6, or $14,050 \times 0.6 = 8430$ lb. Then the factor of safety against sliding is $8430 \div 7900$, or 1.07. This factor of safety is too low, so the profile of the trial section would have to be modified. A projection below the base, similar to that shown in Figure 18-1a, materially increases the resistance to sliding.

19

Friction

||||||||||||||||||||||||||||||||||||||||||

19-1. Development of Friction

When forces act on objects in such a way as to tend to cause one object to slide on the surface of another object, a resistance to the sliding motion is often developed at the contact face between the objects. This resistance is called *friction*, and it constitutes a special kind of force.

For the object shown in Figure 19-1a, being acted on by its own weight and the inclined force F, we may observe that the motion that is impending is one of sliding of the block toward the right along the surface of the plane. The force that tends to cause this motion is the component of F that is parallel to the plane. The component of F that is vertical works with the weight of the block to press the block against the plane. The sum of the weight plus the vertical component of F (F sin Θ) is called the pressure on the plane or the force normal (perpendicular) to the plane.

A free-body diagram of the block is shown in Figure 19-1b. For equilibrium of the block, two components of resistance must be developed. For equilibrium in a direction normal to the plane, the reactive force N is required, its magnitude being equal to $W + F$ sin Θ. For equilibrium in a horizontal direction, along the surface of the plane, a frictional resistance must be developed with a magnitude equal to F cos Θ.

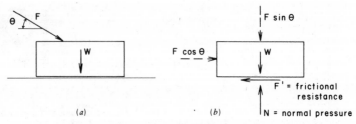

FIGURE 19-1.   Development of sliding friction.

The situation just described may result in one of three possibilities, as follows:

1. The block does not move because the potential frictional resistance is more than adequate, that is,

$$F' > F \cos \Theta$$

2. The block moves because the friction is not of sufficient magnitude, that is,

$$F' < F \cos \Theta$$

3. The block is just on the verge of moving because the potential friction is exactly equal to the force tending to induce sliding, that is,

$$F' = F \cos \Theta$$

From observations and experimentation the following deductions have been made about friction.

1. The frictional resisting force (F' in Figure 19-1b) always opposes motion; that is, it acts opposite to the slide-inducing force.
2. For dry, smooth surfaces, the frictional resistance developed up to the point of motion is directly proportional to the normal pressure between the surfaces. We thus express the maximum value for the frictional resistance as

$$F' = \mu N$$

in which μ (Greek lower case mu) is called the *coefficient of friction*.

3. The frictional resistance is independent of the area of contact.

4. The coefficient of static friction (before motion occurs) is greater than the coefficient of kinetic friction (during actual sliding). That is, for the same amount of normal pressure, the frictional resistance is reduced once motion actually occurs.

19-2.   The Coefficient of Friction

Frictional resistance is ordinarily expressed in terms of its maximum potential value. Coefficients for static friction are determined by finding the ratio between the sliding force and the normal pressure at the moment motion just occurs. Simple experiments consist of using a block on an inclined plance with the angle of the plane's slope slowly increased until sliding occurs. (See Figure 19-2a.) Referring to the free-body diagram for the block (Figure 19-2b) we note

$$F' = \mu N = W \sin \phi$$

$$N = W \cos \phi$$

and, as previously noted, the coefficient of friction is expressed as the ratio of F' to N, or

$$\mu = \frac{F'}{N} = \frac{W \sin \phi}{W \cos \phi} = \tan \phi$$

Approximate values of the coefficient of static friction for various combinations of joined objects are given in Table 19-1.

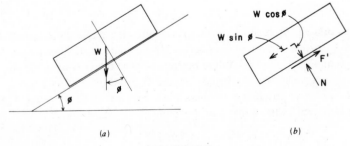

(a)   (b)

FIGURE 19-2

TABLE 19-1. Range of Values for Coefficient of Static Friction

Contact surfaces	Coefficient μ
Wood on wood	0.40–0.70
Metal on wood	0.20–0.65
Metal on metal	0.15–0.30
Metal on stone, masonry, concrete	0.30–0.70

19-3. Solving Problems with Friction

Problems involving friction are usually one of two types. The first involves situations in which friction is one of the forces in a system, and the problem is to determine whether the frictional resistance is sufficient to maintain the equilibrium of the system. For this type of problem the solution consists of writing the equations for equilibrium, including the maximum potential friction $F' = \mu N$, and interpreting the results. If the frictional resistance is not large enough, sliding will occur; if it is just large enough or excessive, sliding will not occur.

The second type of problem involves situations in which the force required to overcome friction is to be found. In this case the slide-inducing force is simply equated to the frictional resistance, and the required force is determined.

Note that in these problems sliding motions are usually of the form of pure translation (no rotation), and we may therefore treat the force systems as simple concurrent ones, ignoring moment effects. An exception is the problem in which tipping is possible, shown in Example 4.

Example 1. A block is placed on an inclined plane whose angle is slowly increased until sliding occurs. If the angle of the plane with the horizontal is 35° when sliding occurs, what is the coefficient of static friction between the block and the plane? (See Figure 19-3.)

Solution: As previously derived, the coefficient of friction may simply be stated as the tangent of the angle of the plane. Thus

$$\mu = \tan \phi = \tan 35° = 0.70$$

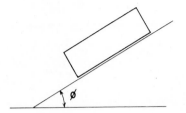

FIGURE 19-3

Example 2. Find the horizontal force P required to slide a block weighing 100 lb if the coefficient of static friction is 0.30. (See Figure 19-4.)

Solution: For sliding to occur, the slide-inducing force P must be slightly larger than the frictional resistance F'. Equating P to F'

$$P = F' = \mu N = 0.30(100) = 30 \text{ lb} \pm$$

The force must be slightly larger than 30 lb.

Example 3. A block is pressed against a vertical wall with a force of 20 lb which acts upward at an angle of 30° with the horizontal. (See Figure 19-5a.)

(a) Express the frictional resistance to motion in terms of the available pressure.

(b) If the block weighs 15 lb and the coefficient of static friction is 0.40, will the block slide?

(c) At what angle must the 20-lb force act to cause the 15-lb block to slide upward if the coefficient of static friction is 0.40?

Solution: For (a):

$$F' = \mu N = \mu(20 \cos 30°) = 17.32\mu \text{ lb}$$

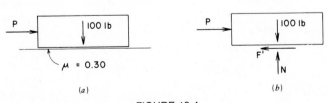

(a)   (b)

FIGURE 19-4

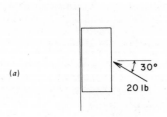

(a)

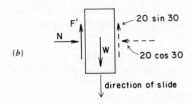

(b)

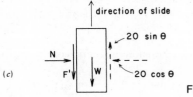

(c)

FIGURE 19-5

For (b):

$$\text{sliding resistance} = \text{net sliding force}$$

$$F' = (W) - (20 \sin 30°)$$

$$= W - 10$$

Then, from (a)

$$(17.32)(0.40) = 15 - 10$$

$$6.93 > 5, \text{ so sliding will not occur}$$

For (c):

$$\text{sliding resistance} = \text{net sliding force}$$

$$F' = 20 \sin \Theta - 15$$

$$0.40(20 \cos \Theta) = 20 \sin \Theta - 15$$

$$20 \sin \Theta - 8 \cos \Theta = 15$$

$$\frac{20 \sin \Theta}{8 \cos \Theta} - 1 = 15$$

$$\frac{\sin \Theta}{\cos \Theta} = \tan \Theta = \frac{8}{20}\,(16) = 6.4$$

$$\Theta = \tan^{-1} 6.4 = 81.1°$$

Example 4. Will the block shown in Figure 19-6a tip or slide?
Solution: We first determine whether the block will slide by computing the maximum frictional resistance and comparing it to the slide-
inducing force.

$$\text{Maximum } F' = 0.60(8) = 4.8 \text{ lb}$$

and therefore the block will not slide.

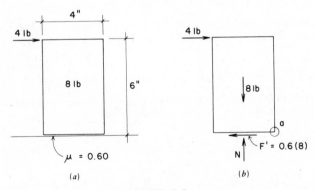

FIGURE 19-6

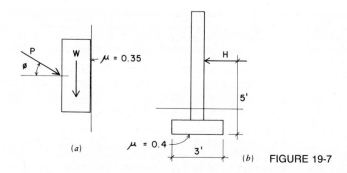

(a)   (b)   FIGURE 19-7

We then evaluate the tipping potential by considering the block to be horizontally restrained at the base by the frictional resistance. Tipping will be considered in terms of overturning about the lower corner (*a* in Figure 19-6*b*). The overturning moment is

$$M = 4 \times 6 = 24 \text{ in.-lb}$$

Resisting this is the moment of the block's weight which is

$$M = 8 \times 2 = 16 \text{ in.-lb}$$

Since the stabilizing moment offered by the weight is not sufficient, the block will tip.

Problem 19-3-A.   Find the angle at which the block shown in Figure 19-2 will slip if the coefficient of friction is 0.35.

Problem 19-3-B.   For the block shown in Figure 19-7*a* find the value of P required to keep the block from slipping if $\phi = 10°$ and $W = 10$ lb.

Problem 19-3-C*.   For the block shown in Figure 19-7*a*, find the weight for the block that will result in slipping if $\phi = 15°$ and $P = 10$ lb.

Problem 19-3-D*.   For the wall and footing shown in Figure 19-7*b*, make separate determinations of the possibilities for slipping and overturn. $H = 1000$ lb and $W = 3000$ lb.

20

Combined Stresses
‖‖

The work presented thus far in this book has involved only singular stress conditions; that is, stress produced by a single phenomenon such as tension only, bending only, and so on. The beam, for example, has simultaneous conditions of direct stress (tension and compression) due to bending and shear stress due to the internal shear force. We have considered the stresses due to bending and shear in the beam as separate events, but have yet to consider their actual combinations. The material in this chapter deals with this and some other common stress combinations.

20-1. Combined Axial Force and Bending

Various situations occur in which both axial force of tension or compression and a bending moment exist at the same cross section in a structural member. Consider the hanger shown in Figure 20-1, in which a 2-in. square steel bar is welded to a plate which is bolted to the bottom of a wood beam. A short piece of steel with a hole in it is welded to the face of the bar, and a load is hung by some means from the hole. In this situation the steel bar is subjected to combined actions of tension and bending, both of which are produced by the hung load. The bending occurs because the load is not applied axially to the bar; the bending moment thus produced has a magnitude of $2 \times 5000 = 10,000$ in.-lb [$50 \times 22 = 1100$ kN-mm, or 1.10 kN-m].

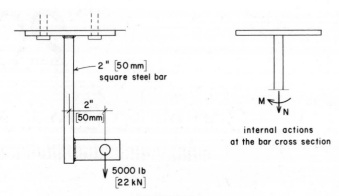

FIGURE 20-1

For this simple case, the stresses due to the two phenomena are found separately and added as follows. For the direct tension effect

$$f_a = \frac{N}{A} = \frac{5000}{4} = 1250 \text{ psi } [8.8 \text{ MPa}]$$

For the bending we first find the section modulus of the cross section, thus

$$S = \frac{bd^2}{6} = \frac{2(2)^2}{6} = 1.333 \text{ in.}^3 \, [20.83 \times 10^3 \text{ mm}^3]$$

Then, for the bending stress

$$f_b = \frac{M}{S} = \frac{10,000}{1.333} = 7502 \text{ psi } [52.8 \text{ MPa}]$$

and the stress combinations are

maximum $f = f_a + f_b = 1250 + 7502 = 8752$ psi [61.6 MPa]

minimum $f = f_a - f_b = 1250 - 7502 = 6252$ psi [44.0 MPa]

Note that although the tension stress due to the direct force is uniformly distributed on the cross section, the bending stress is not. The combined condition is therefore as shown in Figure 20-2c, and the maximum and minimum stresses just computed are the edge stresses as shown in the figure. For the algebraic work the tension stress was considered positive, which means that compressive stress is then neg-

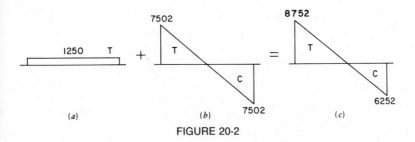

FIGURE 20-2

ative. It is necessary to be aware of this algebraic sign condition when combining stresses.

Figure 20-3 shows a situation in which a simple rectangular footing is subjected to a combination of forces that require the resistance of vertical force, horizontal sliding, and overturning moment. The development of resistance to horizontal movement is produced by some combination of friction on the bottom of the footing and horizontal earth pressure on the face of the footing. Our concern here is for the investigation of the vertical force and the overturning moment and the resulting combination of vertical soil pressures that they develop.

Figure 20-3 illustrates our usual approach to the combined direct force and moment on a cross section. In this case the "cross section" is the contact face of the footing with the soil. However the combined force and moment may originate, we make a transformation into an equivalent eccentric force that produces the same effects on the cross section. The direction and magnitude of this mythical equivalent e are related to properties of the cross section in order to qualify the nature of the stress combination. The value of e is established by simply dividing the force normal to the cross section by the moment, as shown in the figure. The net, or combined, stress distribution on the section is visualized as the sum of the separate stresses due to the normal force and the moment. For the stresses on the two extreme edges of the footing the general formula for the combined stress is

$$p = \frac{N}{A} \pm \frac{Nec}{I}$$

We observe three cases for the stress combination obtained from this formula, as shown in the figure. The first case occurs when e is small, resulting in very little bending stress. The section is thus sub-

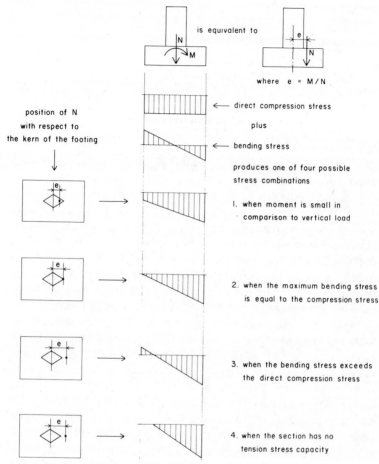

FIGURE 20-3.   Analysis for stress due to combined compression and moment.

jected to all compressive stress, varying from a maximum value on one edge to a minimum on the opposite edge.

The second case occurs when the two stress components are equal, so that the minimum stress becomes zero. This is the boundary condition between the first and third cases, since any increase in the eccentricity will tend to produce some tension stress on the section. This is a significant limit for the footing since tension stress is not possible for the soil-to-footing contact face. Thus Case 3 is possible only in a beam or column where tension stress can be developed. The value of e that corresponds to Case 2 can be derived by equating the two components of the stress formula as follows:

$$\frac{N}{A} = \frac{Nec}{I}, \quad e = \frac{I}{Ac}$$

This value for e establishes what is called the kern limit of the section. The kern is a zone around the centroid of the section within which an eccentric force will not cause tension on the section. The form of this zone may be established for any shape of cross section by application of the formula derived for the kern limit. The forms of the kern zones for three common shapes of section are shown in Figure 20-4.

When tension stress is not possible, eccentricities beyond the kern limit will produce a so-called cracked section, which is shown as Case 4 in Figure 20-3. In this situation some portion of the section becomes unstressed, or cracked, and the compressive stress on the remainder of the section must develop the entire resistance to the force and moment.

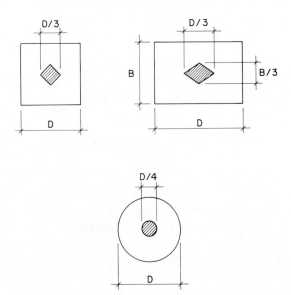

FIGURE 20-4.   Kern limits for common shapes.

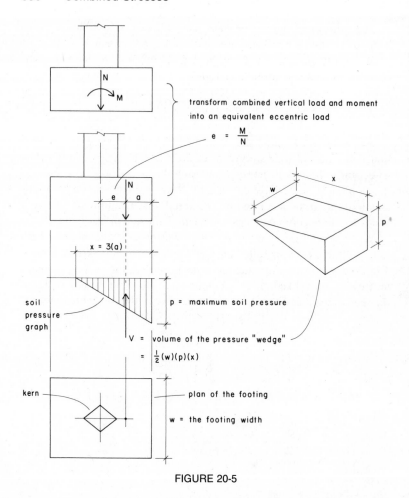

transform combined vertical load and moment
into an equivalent eccentric load

$$e = \frac{M}{N}$$

x = 3(a)

p = maximum soil pressure

V = volume of the pressure "wedge"

$$= \frac{1}{2}(w)(p)(x)$$

soil
pressure
graph

kern

plan of the footing

w = the footing width

FIGURE 20-5

Figure 20-5 shows a technique for the analysis of the cracked section, called the pressure wedge method. The pressure wedge represents the total compressive force developed by the soil pressure. Analysis of the static equilibrium of this wedge and the force and moment on the section produces two relationships that may be utilized to establish the dimensions of the stress wedge. These relationships are as follows:

1. The total volume of the wedge is equal to the vertical force on the section. (Sum of the vertical forces equals zero.)

2. The centroid of the wedge is located on a vertical line with the
    force on the section. (Sum of the moments on the section equals
    zero.)

Referring to Figure 20-5, the three dimensions of the stress wedge are
w, the width of the footing, p, the maximum soil pressure, and x, the
limit of the uncracked portion of the section. With w known, the solution of the wedge analysis consists of determining values for p and
x. For the rectangular footing, the simple triangular stress wedge will
have its centroid at the third point of the triangle. As shown in the
figure, this means that x will be three times the dimension a. With the
value for e determined, a may be found and the value of x established.

The volume of the stress wedge may be expressed in terms of its
three dimensions as follows:

$$V = \tfrac{1}{2}wpx$$

Using the static equilibrium relationship stated previously, this volume may be equated to the force on the section. Then, with the values
of w and x established, the value for p may be found as follows:

$$N = V = \tfrac{1}{2}wpx$$

$$p = \frac{2N}{wx}$$

Example.   Find the maximum value of the soil pressure for the footing shown in Figure 20-6. The compression force at the bottom of the
footing N is 100 k [450 kN], and the moment is 100 k-ft [135 kN-m].
Find the pressure for footing widths of (1) 8 ft, (2) 6 ft, and (3) 5 ft.
Solution: The first step is to determine the equivalent eccentricity and
compare it to the kern limit for the footing to establish which of the

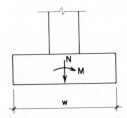

FIGURE 20-6

cases shown in Figure 20-3 applies. We thus compute for (1)

$$e = \frac{M}{N} = \frac{100}{100} = 1 \text{ ft } [0.3 \text{ m}]$$

kern for the 8-ft wide footing $= \frac{8}{6} = 1.33$ ft [0.41 m]

and it is established that Case 1 applies.

We next determine the soil pressure, using the formula for the combined stress as previously derived.

$$p = \frac{N}{A} + \frac{Mc}{I} = \frac{100}{64} + \frac{100 \times 4}{341.3} = 1.56 + 1.17 = 2.73 \text{ ksf}$$

$$[75.6 + 56.1 = 131.7 \text{ kPa}]$$

in which

$$A = (8)^2 = 64 \text{ ft}^2 \; [5.95 \text{ m}^2]$$

$$I = \frac{bd^3}{12} = \frac{(8)^4}{12} = 341.3 \text{ ft}^4 \; [2.95 \text{ m}^4]$$

For (2) it may be observed that the kern limit is $\frac{6}{6} = 1$, which is equal to the eccentricity. Thus the situation is that shown as Case 2 in Figure 20-3, and the pressure is such that $N/A = Mc/I$. Thus

$$p = 2\left(\frac{N}{A}\right) = 2\left(\frac{100}{6 \times 6}\right) = 5.56 \text{ ksf } [266 \text{ kPa}]$$

For (3) the eccentricity exceeds the kern limit, and the investigation must be done as illustrated in Figure 20-5.

$$a = \frac{5}{2} - e = 2.5 - 1 = 1.5 \text{ ft } [0.76 - 0.3 = 0.46 \text{ m}]$$

$$x = 3a = 3(1.5) = 4.5 \text{ ft } [1.38 \text{ m}]$$

$$p = \frac{2N}{wx} = \frac{2(100)}{5 \times 4.5} = 8.89 \text{ ksf } [429 \text{ kPa}]$$

Problem 20-1-A. For the hanger rod shown in Figure 20-7 find the maximum and the minimum values of the tension stress (or the net compressive stress).

Problem 20-1-B. Find the value of the maximum soil pressure for a footing such as

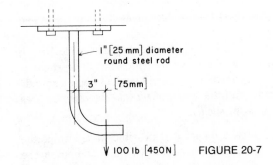

1" [25 mm] diameter
round steel rod

3"   [75mm]

100 lb [450N]   FIGURE 20-7

that shown in Figure 20-6. The compression force is 150 k [670 kN], and the moment is 175 k-ft [240 kN-m]. Find soil pressures for footings with side dimensions of (1) 9 ft [2.74 m], (2) 7 ft [2.13 m], and (3) 6 ft [1.83 m].

20-2.   Development of Shear Stress

Shear force generates a lateral, slicing effect in materials. Visualized in two dimensions, this direct effect is as shown in Figure 20-8a. For stability within the material, there will be a counteracting, or reactive, shear stress developed at a right angle to the active stress, as shown in Figure 20-8b. Finally, the interaction of the direct and reactive shears will produce both diagonal tension and diagonal compression stresses, as shown in Figure 20-8c and d. In concrete the critical stress concern is for the diagonal tension stress, which works on the weakest property of the material.

Referring to Figure 20-8, we may observe that:

1.   The unit reactive (right angle) shear stress will be equal in magnitude to the unit active shear stress.
2.   The combined diagonal effect (tension or compression) will be

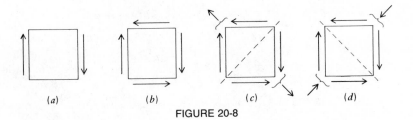

(a)   (b)   (c)   (d)

FIGURE 20-8

$\sqrt{2}$ times the shear effect (vector combination of the direct and reactive shears).

3. The diagonal stress will develop on the diagonal plane, which is $\sqrt{2}$ times the size of the plane on which the shear stress is developed. Thus the diagonal stress will be the same magnitude as the shear stress; both the force effect and the stressed area being $\sqrt{2}$ times that for shear.

Accepting the observations just made, it is possible to determine the critical tension stress that accompanies shear by simply computing the unit shear stress. As in other situations, however, it is also necessary to determine the *direction* of the tension stress, especially when reinforcement must be provided.

20-3.   Stress on an Oblique Cross Section

Just as shear was shown to produce direct stresses, so we may show that direct force produces shear stress. Consider the element shown in Figure 20-9, subjected to a tension force. If a section is cut that is not at a right angle to the force, but rather is at some angle θ to it, there may be seen to exist two components of the internal force P. One component is at a right angle to the cut section surface plane and the other is in the surface plane. These two components produce, respectively, direct tension stress and shear stress at the cut section. We may

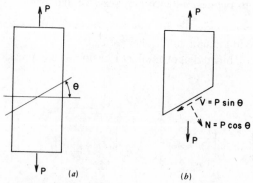

FIGURE 20-9.   Stress on an oblique cross section.

express these stresses as follows:

$$f = \frac{P \cos \Theta}{A/\cos \Theta} = \frac{P}{A} \cos^2 \Theta$$

$$v = \frac{P \sin \Theta}{A/\cos \Theta} = \frac{P}{A} \sin \Theta \cos \Theta$$

We note that when $\Theta = 0$, $\cos \Theta = 1$ and $\sin \Theta = 0$, and therefore $f = P/A$ and $v = 0$. Also, when $\Theta = 45°$, $\cos \Theta = \sin \Theta = (2)^{1/2}/2$, and therefore $f = \frac{1}{2}(P/A)$ and $v = \frac{1}{2}(P/A)$.

Example. The wood block shown in Figure 20-10a has its grain at an angle of 30° to the direction of force. Find the compression and shear stresses on a plane parallel to the grain.
Solution: Note that as used in Figure 20-9, $\Theta = 60°$. Then for the free-body diagram shown in Figure 20-10b

$$N = P \cos 60°, \quad V = P \sin 60°, \quad A = (4 \times 3)/\cos 60°$$

The forces and the area may now be used directly to find the stresses. Or we may use the formulas derived in the preceding work.

$$f = \frac{P}{A} \cos^2 \Theta = \frac{1200}{12} (0.5)^2 = 25 \text{ psi } [180 \text{ kPa}]$$

$$v = \frac{P}{A} \sin \Theta \cos \Theta = \frac{1200}{12} (0.5)(0.866) = 43.3 \text{ psi } [312 \text{ kPa}]$$

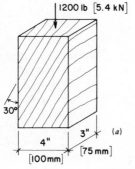

FIGURE 20-10

Problem 20-3-A-B*-C. A structural member is loaded in compression as shown in Figure 20-9. Find the value of the direct and shear stresses if the compression force P is 10,000 lb [45 kN], the area of the cross section is 10 in.^{2} [6450 mm^2], and the cross section is at an angle to the load direction with Θ (as shown in Figure 20-9) equal to: (A) 15°, (B) 20°, (C) 30°.

20-4.   Combined Direct and Shear Stresses

The stress actions shown in Figure 20-8 represent the conditions that occur when shear alone is considered. When shear occurs simultaneously with other effects, the various resulting stress conditions must be combined to produce the net effect. Figure 20-11 shows the result of combining a shear stress effect with a direct tension stress effect. For shear alone, the critical tension stress plane is at 45°, as shown in Figure 20-8a. For the tension alone, the critical tension stress plane is at 90°. For the combined stress condition, the unit stress will be some magnitude higher than either the shear or direct tension stress, and the critical tension stress plane will be at an angle somewhere between 45 and 90°.

Consider the beam shown in Figure 20-12. Various combinations of direct and shear stresses may be visualized in terms of the conditions at the cross section labeled *S-S* in the figure. With reference to the points on the section labeled 1 through 5, we observe the following:

At point 1 the beam shear stress is zero, and the dominant stress is the compressive bending stress in a horizontal direction.

At point 3 the shear stress is maximum, bending stress is zero, and the tension stress due to shear is at an angle of 45°.

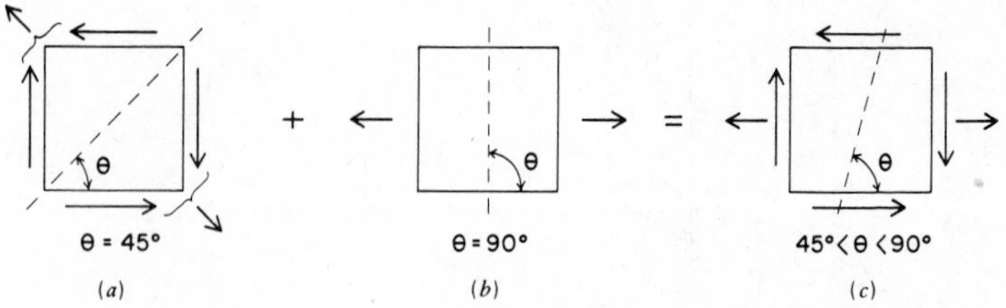

FIGURE 20-11

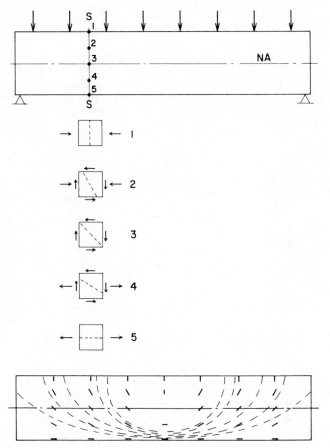

FIGURE 20-12.   Development of net tension in a beam.

At point 5 the shear stress is zero; bending stress in tension predominates, acting in a horizontal direction.

At point 4 the net tension stress due to the combination of shear and direct tension force will operate in a direction somewhere between the horizontal and 45°.

At point 2 the net tension will operate at an angle somewhat larger than 45°.

At point 1 the tension goes to zero as the angle of its direction approaches 90°.

The direction of the net tension stress is indicated for various points in the beam by the short dark bars on the beam elevation at the bottom of Figure 20-12. The dashed lines indicate the flow of tension. If this figure were inverted, it would indicate the action of the net compressive stresses in the beam.

21

Noncoplanar Force Systems

||||||||||||||||||||||||||||||||||||||||||||||

Forces and structures exist in reality in a three-dimensional world. The work in preceding chapters has been limited mostly to systems of forces operating in two-dimensional planes. This is commonly done in practice, primarily for the same reasons that we have done it here: it makes both visualization and computations easier. As long as the full three-dimensional character of the forces and the structure are eventually dealt with, this approach is usually quite adequate. For visualization, as well as some computations, however, it is sometimes necessary to work with forces in noncoplanar systems. This chapter presents some exercises that will help in the development of an awareness of the problems of working with such force systems.

Graphical representation, visualization, and any mathematical computation all become more complex with noncoplanar systems. The following discussions rely heavily on the examples to illustrate basic concepts. The orthogonal axis system $x - y - z$ is used for ease of both visualization and computation.

Units of measurement for both forces and dimensions are of small significance in this work. Because of this, and because of the complexity of both the graphical representations and the mathematical computations, the conversions for SI units have been omitted, except for the exercise problems and answers.

21-1.   Noncoplanar, Concurrent, Nonparallel Systems

Figure 21-1 shows a single force acting in such a manner that it has
component actions in three dimensions. That is, it has x, y, and z components. If this force represents the resultant of a system of forces, it
may be identified as follows:

$$R = \sqrt{(\Sigma F_x)^2 + (\Sigma F_y)^2 + (\Sigma F_z)^2}$$

$$\cos \theta_x = \frac{\Sigma F_x}{R}, \qquad \cos \theta_y = \frac{\Sigma F_y}{R}, \qquad \cos \theta_z = \frac{\Sigma F_z}{R}$$

Equilibrium for this type of system can be established by fulfilling the
following conditions:

$$\Sigma F_x = 0, \qquad \Sigma F_y = 0, \qquad \Sigma F_z = 0$$

Example 1.   Find the resultant of the three forces shown in Figure
21-2a.

Solution: The technique used is that of solving for the geometry of
the force lines first, and then using proportionate relationships to find
the force vector components. Although trigonometry could also be
used, the direct use of simple geometry makes the visualization of the
actions easier.

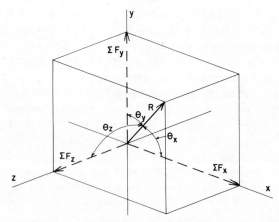

FIGURE 21-1.   Components of noncoplanar forces.

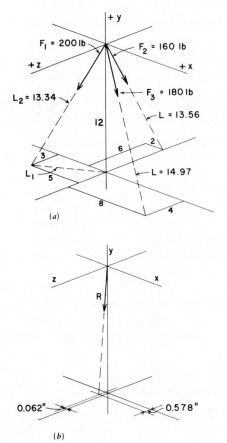

(a)

(b)   FIGURE 21-2

Referring to the line lengths shown in Figure 21-2a:

$$L_1 = \sqrt{(5)^2 + (3)^2} = \sqrt{34} = 5.83$$

$$L_2 = \sqrt{(12)^2 + (34)} = \sqrt{178} = 13.34$$

Note: To reinforce the point that the unit of measurement for dimensions is not relevant, we have omitted it.

The other line lengths can be established in a similar manner. Their values are shown on the figure. The determination of the force components and their summation is presented in Table 21-1. Note that the

TABLE 21-1. Summation of Force Components (Figure 21-2)

Force	x component	y component	z component
F_1	$200\,\dfrac{5}{13.34} = 75 \nwarrow$	$200\,\dfrac{12}{13.34} = 180\downarrow$	$200\,\dfrac{3}{13.34} = 45\swarrow$
F_2	$160\,\dfrac{2}{13.56} = 23.6 \nwarrow$	$160\,\dfrac{12}{13.56} = 141.7\downarrow$	$160\,\dfrac{6}{13.56} = 70.8\nearrow$
F_3	$180\,\dfrac{8}{14.97} = 96.2 \searrow$	$180\,\dfrac{12}{14.97} = 144.4\downarrow$	$180\,\dfrac{4}{14.97} = 48.2\swarrow$
	$\Sigma F_x = 2.4\,\text{lb}\nwarrow$	$\Sigma F_y = 466.1\,\text{lb}\downarrow$	$\Sigma F_z = 22.4\,\text{lb}\swarrow$

signs of the components (+ and −) are established with reference to the + directions indicated for the three axes, as shown in Figure 21-2a. Using the summations, we determine the value of the resultant as

$$R = \sqrt{(2.4)^2 + (466.1)^2 + (22.4)^2} = \sqrt{217{,}757} = 466.6$$

The direction of R may be established by expressing the three cosine equations, as described earlier, or by establishing its point of intersection with the x-z plane, as shown in Figure 21-2b. Using the latter method,

$$\frac{\Sigma F_x}{\Sigma F_y} = \frac{x \text{ distance from } z\text{-axis}}{12} = \frac{2.4}{466.1}$$

therefore

$$x \text{ distance from } z\text{-axis} = \frac{2.4}{466.1}\,(12) = 0.062$$

and, similarly

$$z \text{ distance from } x\text{-axis} = \frac{22.4}{466.1}\,(12) = 0.578$$

Example 2. For the structure shown in Figure 21-3a, find the tension in the guy wires and the compression in the mast for the loading indicated.

Solution: As in Example 1, the geometry of the wires is established

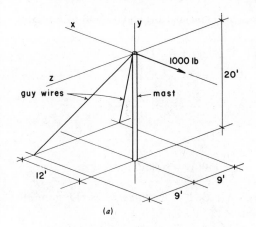

(a)

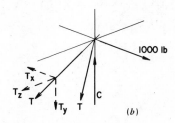

(b)   FIGURE 21-3

first. Thus

$$L = \sqrt{(9)^2 + (12)^2 + (20)^2} = \sqrt{625} = 25$$

Consider the concentric forces that act at the top of the mast. For equilibrium in the x direction

$$\Sigma F_x = 0 = 1000 - 2(T_x), \qquad T_x = 500 \text{ lb}$$

Then, from the geometry of the wire

$$\frac{T}{T_x} = \frac{25}{12}, \qquad T = \frac{25}{12}(T_x) = \frac{25}{12}(500) = 1041.67 \text{ lb}$$

For the compression in the mast, consider the equilibrium of forces in the y direction. Thus

$$\Sigma F_y = 0 = C - 2(T_y), \qquad C = 2(T_y)$$

where

$$\frac{T_y}{T_x} = \frac{20}{12}, \qquad T_y = \frac{20}{12}(T_x) = \frac{20}{12}(500)$$

thus

$$C = 2 \times \frac{20}{12}(500) = 1666.67 \text{ lb}$$

Example 3.   Find the tension in each of the three wires in Figure 21-4 due to the force indicated.

Solution: As before, we find the lengths of the wires as follows:

$$L_1 = \sqrt{(5)^2 + (4)^2 + (20)^2} = \sqrt{441} = 21$$

$$L_2 = \sqrt{(2)^2 + (8)^2 + (20)^2} = \sqrt{468} = 21.63$$

$$L_3 = \sqrt{(12)^2 + (20)^2} = \sqrt{544} = 23.32$$

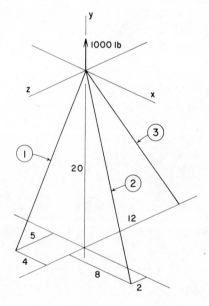

FIGURE 21-4

The three equilibrium equations for the concentric forces are thus

$$\Sigma F_x = 0 = \frac{4}{21} T_1 - \frac{8}{21.63} T_2 + \frac{0}{23.32} T_3$$

$$\Sigma F_z = 0 = \frac{5}{21} T_1 + \frac{2}{21.63} T_2 - \frac{12}{23.32} T_3$$

$$\Sigma F_y = 0 = \frac{20}{21} T_1 + \frac{20}{21.63} T_2 + \frac{20}{23.32} T_3 - 1000$$

Solution of these three equations with three unknowns yields the following:

$$T_1 = 525 \text{ lb}, \qquad T_2 = 271 \text{ lb}, \qquad T_3 = 290 \text{ lb}$$

Problem 21-1-A. Find the resultant of the three forces shown in Figure 21-5a. Establish the direction of the resultant by finding the coordinates of its intersection with the x-z plane.

Problem 21-1-B*. Find the compression force in the struts and the tension force in the wire for the structure in Figure 21-5b.

Problem 21-1-C. Find the tension in each of the wires for the system shown in Figure 21-5c.

21-2. Noncoplanar, Nonconcurrent, Parallel Systems

Consider the force system shown in Figure 21-6. Assuming the direction of the forces to be parallel to the y-axis, the resultant can be stated as

$$R = \Sigma F_y$$

and its location in the x-z plane can be established (see Figure 21-6) by two moment equations, taken with respect to the x-axis and the z-axis; thus

$$L_x = \frac{\Sigma M_z}{R} \quad \text{and} \quad L_z = \frac{\Sigma M_x}{R}$$

Equilibrium for the system can be established by fulfilling the following conditions:

$$\Sigma F_y = 0, \qquad \Sigma M_x = 0, \qquad \Sigma M_z = 0$$

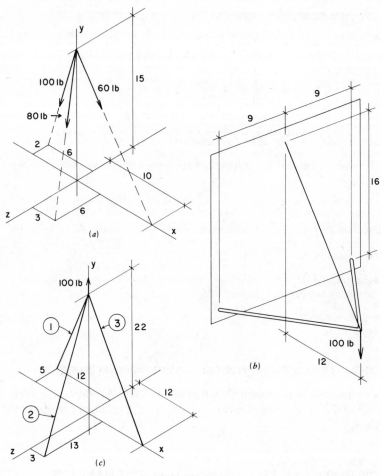

FIGURE 21-5

As with the coplanar parallel systems, the resultant may be a couple. That is, the summation of forces may be zero, but there may be a net rotational effect about the x-axis and/or the z-axis. When this is the case, the resultant couple may be visualized in terms of two component couples, one in the x-y plane (for ΣM_z) and one in the z-y plane (for ΣM_x). See Example 2 in the following work.

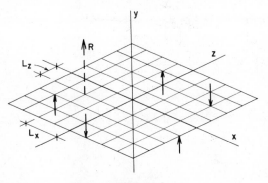

FIGURE 21-6

Example 1. Find the resultant of the system shown in Figure 21-7.
Solution: We first find the resultant force; thus

$$R = \Sigma F_y = 50 + 60 + 160 + 80 = 350 \text{ lb}$$

Then, for its location:

$$\Sigma M_x = +(160 \times 8) - (60 \times 6) = 920 \text{ ft-lb}$$

$$\Sigma M_z = +(50 \times 8) - (80 \times 15) = 800 \text{ ft-lb}$$

and the distances from the axes are

$$L_x = \frac{800}{350} = 2.29 \text{ ft}, \qquad L_z = \frac{920}{350} = 2.63 \text{ ft}$$

Example 2. Find the resultant of the system shown in Figure 21-8.

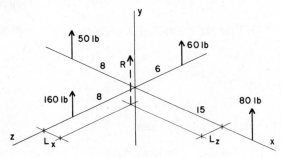

FIGURE 21-7

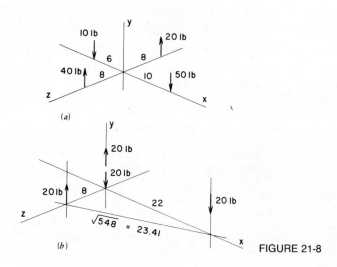

FIGURE 21-8

Solution:

$$R = F_y = +40 + 20 - 10 - 50 = 0$$

$$\Sigma M_x = +(40 \times 8) - (20 \times 8) = 160 \text{ ft-lb}$$

$$\Sigma M_z = +(10 \times 6) - (50 \times 10) = 440 \text{ ft-lb}$$

The resultant is seen to be a couple with the two components just determined. If necessary, these two components can be combined into a single-force couple, although it may be sufficient to use the components for some problems.

Example 3. Find the tension in the three wires in the system shown in Figure 21-9.

Solution: Using the three equilibrium conditions:

$$\Sigma F_y = 0 = T_1 + T_2 + T_3 - 100$$

$$\Sigma M_x = 0 = 4T_1 - 6T_2$$

$$\Sigma M_z = 0 = 6T_1 - 8T_3$$

Solution of these simultaneous equations yields

$$T_1 = 41.4 \text{ lb}, \qquad T_2 = 27.6 \text{ lb}, \qquad T_3 = 31.0 \text{ lb}$$

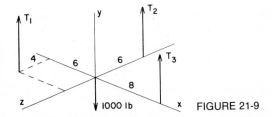

FIGURE 21-9

Problem 21-2-A.   Find the resultant and its location with respect to the *x* and *z* axes for the system shown in Figure 21-10*a*.

Problem 21-2-B.   Find the tension in the three wires of the system shown in Figure 21-10*b*.

21-3.   Noncoplanar, Nonconcurrent, Nonparallel System

This is the general spatial force system with no simplifying conditions regarding its geometry. The resultant for such a system may be any of four possibilities, as follows:

1.   Zero—if the system is in equilibrium
2.   A force—if *F* is not zero
3.   A couple—if *M* is not zero
4.   A force plus a couple—which is the general case

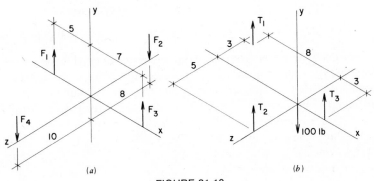

FIGURE 21-10

If the resultant is a force, its magnitude is determined as

$$R = \sqrt{(\Sigma F_x)^2 + (\Sigma F_y)^2 + (\Sigma F_z)^2}$$

and its direction by

$$\cos \theta_x = \frac{\Sigma F_x}{R}, \qquad \cos \theta_y = \frac{\Sigma F_y}{R}, \qquad \cos \theta_z = \frac{\Sigma F_z}{R}$$

If the resultant is a couple, it may be determined in terms of its component moments about the three axes in a procedure similar to that shown for the parallel systems in Section 21-2. For most purposes these component moments will be sufficient, although their geometric combination into a single couple is possible.

Equilibrium for the general spatial force system can be established by fulfilling the following conditions:

$$\Sigma F_x = 0, \qquad \Sigma F_y = 0, \qquad \Sigma F_z = 0$$

The potential complexity of this system makes for a large number of possible situations. Although in theory a problem could involve the necessity for solving nine simultaneous equations, simplifying conditions often reduce the complexity of the mathematical work. The larger problem is often the simple visualization and graphic representation of the force system. The following examples illustrate this type of problem; both include simplifying geometric conditions.

Example 1. Find the resultant of the system shown in Figure 21-11. *Solution:* Using the six summations, we determine

$$\Sigma F_x = 50 \text{ lb}, \qquad \Sigma F_y = 100 \text{ lb}, \qquad \Sigma F_z = 20 \text{ lb}$$

$$\Sigma M_x = 700 \text{ ft-lb}, \qquad \Sigma M_y = 500 \text{ ft-lb}, \qquad \Sigma M_z = 600 \text{ ft-lb}$$

For many purposes, such as finding the effective cranking effort or the bending stresses in the crank, it would be sufficient to state the resultant in these component terms. The true net resultant is a force and couple combination. If really necessary, the force may be found as the vector combination of the three force components, its direction as the three cosine functions, and its location with respect to the y-axis in the x-z plane, as shown in Figure 21-11b. The couple may be determined as a single-force couple in a skewed plane by combining the three component moments. Although the net resultant may be of sig-

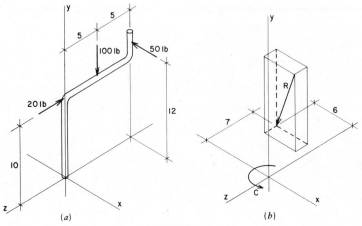

FIGURE 21-11

nificance in machine or navigation problems, it is seldom of direct use in the design of building structures.

The following example more closely resembles a type of problem encountered in structures, involving the establishment of equilibrium and the determination of internal forces in structural members.

Example 2.   Find the tension in the cables and the compression in the boom and mast for the structure shown in Figure 21-12a. Also find the components of the reaction at A.

Solution: Consider the forces acting at the outer end of the boom, as shown in Figure 21-12b.

$$\Sigma F_y = 0 = T'_y - 1000, \qquad T'_y = 1000 \text{ lb}$$

and from the geometry of T'

$$T'_x = \frac{15}{20} T'_y = \frac{15}{20} (1000) = 750 \text{ lb}$$

$$T' = \sqrt{(1000)^2 + (750)^2} = 1250 \text{ lb}$$

Then

$$\Sigma F_x = 0 = C_b - T'_x = C_b - 750, \qquad C_b = 750 \text{ lb}$$

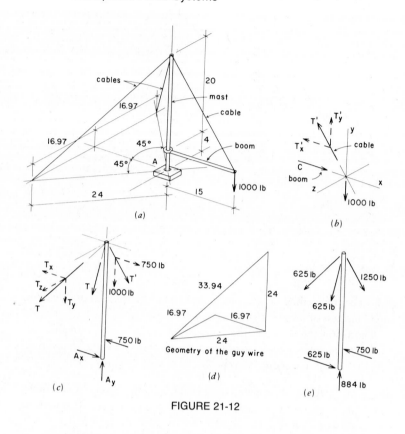

FIGURE 21-12

Next consider the free-body diagram of the mast, as shown in Fig. 21-12c.

$$\Sigma F_x = 0 = A_x + 750 - 2T_x - 750, \qquad A_x = 2T_x$$

$$\Sigma M_z \text{ at } A = 0 = (750 - 2T_x)(24) - (750 \times 4)$$

from which

$$48T_x = 750 \times 20 = 15,000, \qquad T_x = 312.5 \text{ lb}$$

Then, from the geometry of the guy cables, the length of the guy cable is (see Figure 21-12d)

$$L = \sqrt{2} \times 24 = 33.94 \text{ ft}$$

$$T = \frac{33.94}{(\sqrt{2}/2)24}\, T_x = \frac{33.94}{(\sqrt{2}/2)24}\,(312.5) = 625 \text{ lb}$$

And, as previously determined,

$$A_x = 2T_x = 2(312.5) = 625 \text{ lb}$$

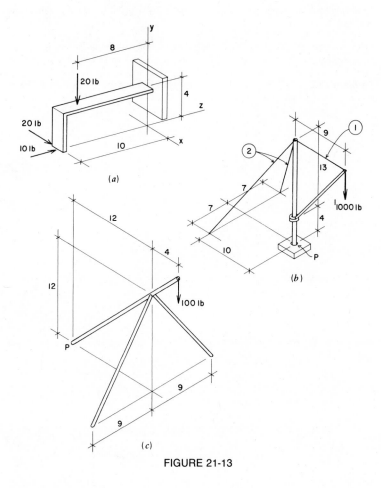

(a)

(b)

(c)

FIGURE 21-13

Then, for the free-body diagram of the mast as shown in Figure 21-12c,

$$\Sigma F_y = 0 = A_y - 2T_y - 1000$$

$$A_y = 2\left(\frac{24}{16.97} \times 312.5\right) + 1000 = 1884 \text{ lb}$$

The results are summarized on Figure 21-12e

Problem 21-3-A.   Find the components of the resultant for the force system shown in Figure 21-13a.

Problem 21-3-B.   Find the tension in the cables, the compression in the boom, and the components of the reaction at P for the structure shown in Figure 21-13b.

Problem 21-3-C*.   Find the compression in the legs and the components of the reaction at P for the structure shown in Figure 21-13c.

22

Use of Internal Conditions
IIIIIIIIIIIIIIIIIIIIIIIIIIIIIIIIIIIIIIIIII

In many structures qualifying conditions exist at supports or within the structure that modify the behavior of the structure, often eliminating some potential components of force actions. Qualification of supports as fixed or pinned has been a situation in most of the structures presented in this work. We now consider some qualification of conditions *within* the structure that modify its behavior.

22-1. Internal Pins

Consider the structure shown in Figure 22-1a. It may be observed that there are four potential components of the reaction forces: A_x, A_y, B_x, and B_y. These are all required for the stability of the structure for the loading shown, and there are thus four unknowns in the investigation of the external forces. Since the loads and reactions constitute a general planar force system, there are three conditions of equilibrium (see Section 2-4; for example: $\Sigma F_x = 0$, $\Sigma F_y = 0$, $\Sigma M_p = 0$). As shown in Figure 22-1a, therefore, the structure is statically indeterminate, not yielding to complete investigation by use of static equilibrium conditions alone.

If the two members of the structure in Figure 22-1 are connected to each other by a pinned joint, as shown at (b), the number of reaction components is not reduced, and the structure is still stable. However,

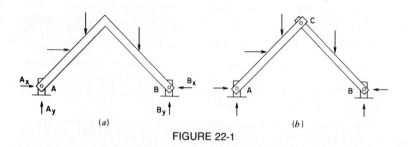

FIGURE 22-1

the internal pin establishes a fourth condition which may be added to the three equilibrium conditions. There are then four conditions that may be used to find the four reaction components. The method of solution for the reactions of this type of structure is illustrated in the following example.

Example.   Find the components of the reactions for the structure shown in Figure 22-2a.

Solution: It is possible to write four equilibrium equations and to solve them simultaneously for the four unknown forces. However, it is always easier to solve these problems if a few tricks are used to simplify the equations. One trick is to write moment equations about points that eliminate some of the unknowns, thus reducing the number of unknowns in a single equation. This was used in the finding of beam reactions in Chapter 7. Consider the free body of the entire structure, as shown in Figure 22-2b.

$$\Sigma M_A = 0 = + (400 \times 5) + (B_x \times 2) - (B_y \times 24)$$

$$24B_y - 2B_x = 2000, \quad \text{or}, \quad 12B_y - B_x = 1000 \qquad (22\text{-}1)$$

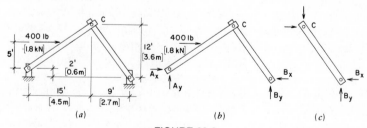

FIGURE 22-2

Now consider the free-body diagram of the right member, as shown in Figure 22-2c.

$$\Sigma M_C = 0 = + (B_x \times 12) - (B_y \times 9)$$

thus

$$B_x = \frac{9}{12} B_y = 0.75 B_y \qquad (22\text{-}2)$$

Substituting Equation (22-2) in Equation (22-1),

$$12B_y - 0.75B_y = 1000, \qquad B_y = \frac{1000}{11.25} = 88.89 \text{ lb}$$

Then, from Equation (22-2),

$$B_x = 0.75B_y = 0.75 \times 88.89 = 66.67$$

Referring again to Figure 22-2b:

$$\Sigma F_x = 0 = A_x + B_x - 400$$

$$0 = A_x + 66.67 - 400$$

$$A_x = 333.33 \text{ lb}$$

$$\Sigma F_y = 0 = A_y + B_y = A_y + 88.89, \qquad A_y = 88.89 \text{ lb}$$

Note that the condition stated in Equation (22-2) is true in this case because the right member behaves as a two-force member. This is not the case if load is directly applied to the member, and the solution of simultaneous equations would be necessary.

Problems 22-1-A-B.   Find the components of the reactions for the structures shown in Figure 22-3.

22-2. Continuous Beams with Internal Pins

The actions of continuous beams are discussed in Chapter 9. It is observed that a beam such as that shown in Fig. 22-4a is statically indeterminate, having a number of reaction components (3) in excess of the conditions of equilibrium for a parallel force system (2). The continuity of such a beam results in the deflected shape and variation of moment as shown beneath the beam in Figure 22-4a. If the beam is

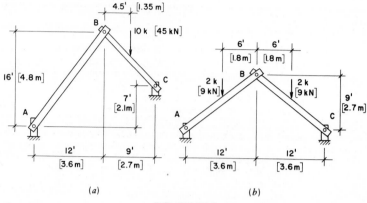

FIGURE 22-3

made discontinuous at the middle support, as shown in Figure 22-4b, the two spans each behave independently as simple beams, with the deflected shapes and moment as shown.

If a multiple-span beam is made internally discontinuous at some point, its behavior may emulate that of a truly continuous beam. For the beam shown in Figure 22-4c the internal pin is located at the point where the continuous beam inflects. Inflection of the deflected shape is an indication of zero moment, and thus the pin does not actually change the continuous nature of the structure. The deflected shape and moment variation for the beam in Figure 22-4c is therefore the same as for the beam in Figure 22-4a. This is true, of course, only for the single loading pattern that results in the inflection point at the same location as the internal pin.

In the first of the following examples the internal pin is deliberately

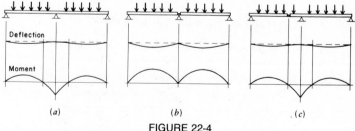

FIGURE 22-4

placed at the point where the beam would inflect if it was continuous. In the second example the pins are placed slightly closer to the support than the location of the natural inflection points. The modification in the second example results in slightly increasing the positive moment in the outer spans, while reducing the negative moments at the supports; thus the values of maximum moment are made closer. If it is desired to use a single-size beam for the entire length, the modification in Example 2 permits design selection of a slightly smaller size.

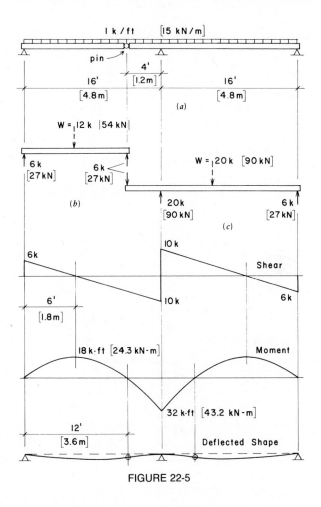

FIGURE 22-5

Example 1. Investigate the beam shown in Figure 22-5*a*. Find the reactions, draw the shear and moment diagrams, and sketch the deflected shape.

Solution: Because of the internal pin, the first 12 ft of the left-hand span acts as a simple beam. Its two reactions are therefore equal, being one-half the total load, and its shear, moment, and deflected shape diagrams are those for a simple beam with a uniformly distributed load. (See Figure 8-10, Case 2.) As shown at (*b*) and (*c*) in Figure 22-5, the

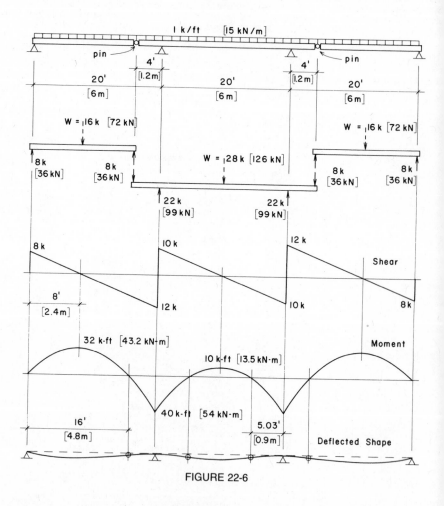

FIGURE 22-6

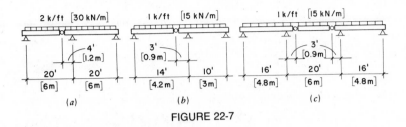

FIGURE 22-7

simple beam reaction at the right end of the 12 ft portion of the left span becomes a 6-kip concentrated load at the left end of the remainder of the beam. This beam (Figure 22-5c) is then investigated as a beam with one overhanging end, carrying a single concentrated load at the cantilvered end and the total distributed load of 20 kips. (Note that on the diagram we indicate the total uniformly distributed load in the form of a single force, representing its resultant.) The second portion of the beam is statically determinate, and we can proceed to find its reactions.

With the reactions known, the shear diagram can be completed. We note the relation between the point of zero shear in the span and the location of maximum positive moment. For this loading the positive moment curve is symmetrical, and thus the location of the zero moment (and beam inflection) is at twice the distance from the end as the point of zero shear. As noted previously, the pin in this example is located exactly at the inflection point of the continuous beam. (For comparison, see Section 9-2.)

Example 2. Investigate the beam shown in Figure 22-6.
Solution: The procedure is essentially the same as for Example 1. Note that this beam with four supports requires two internal pins to become statically determinate. As before, the investigation begins with the consideration of the end portion acting as a simple beam. The second step is to consider the center portion as a beam with two overhanging ends.

Problems 22-2-A-B-C. Investigate the beams shown in Figure 22-7. Find the reactions, draw the shear and moment diagrams, indicating all critical values. Sketch the deflected shape and determine the locations of any inflection points not related to the internal pins. (*Note:* Problem 22-2-C has the same spans and loading as the example in Section 9-3.)

23

Rigid Frames
||||||||||||||||||||||||||||||||||||||||||||||||

Frames in which two or more of the members are attached to each other with connections that are capable of transmitting bending between the ends of the members are called *rigid frames*. The connections used to achieve such a frame are called *moment connections*. Most rigid frame structures are statically indeterminate and do not yield to investigation by consideration of static equilibrium alone. The examples presented in this chapter are all rigid frames that have conditions that make them statically determinate and thus capable of being fully investigated by methods developed in this book.

23-1. Cantilever Frames

Consider the frame shown in Figure 23-1a, consisting of two members rigidly joined at their intersection. The vertical member is fixed at its base, providing the necessary support condition for stability of the frame. The horizontal member is loaded with a uniformly distributed loading and functions as a simple cantilever beam (see Section 8-10). The frame is described as a cantilever frame because of the single fixed support. The five sets of figures shown in Figure 23-1b through f are useful elements for the investigation of the behavior of the frame:

1. The free-body diagram of the entire frame, showing the loads and the components of the reactions (Figure 23-1b). Study of

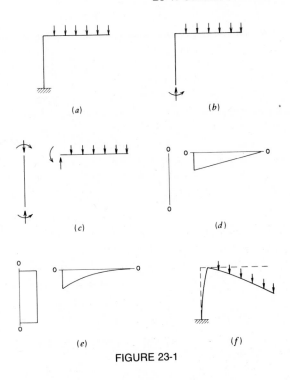

FIGURE 23-1

this figure will help in establishing the nature of the reactions and in the determination of the conditions necessary for stability of the frame.

2. The free-body diagrams of the individual elements (Figure 23-1c). These are of great value in visualizing the interaction of the parts of the frame. They are also useful in the computations for the internal forces in the frame.

3. The shear diagrams of the individual elements (Figure 23-1d). These are sometimes useful for visualizing, or for actually computing, the variations of moment in the individual elements. No particular sign convention is necessary unless in conformity with the sign used for moment. (See discussion of relation of shear and moment in Chapter 8.) Although good as exercises in visualization, the shear diagrams have limited value in the investigation in most cases.

4. The moment diagrams for the individual elements (Figure 23-1e). These are very useful, especially in determination of the deformation of the frame. The sign convention used is that of plotting the moment on the compression side of the element.

5. The deformed shape of the loaded frame (Figure 23-1f). This is the exaggerated profile of the bent frame, usually superimposed on an outline of the unloaded frame for reference. This is very useful for the general visualization of the frame behavior. It is particularly useful for determination of the character of the external reactions and the form of interaction between the parts of the frame.

When performing investigations, these elements are not usually produced in the sequence just described. In fact, it is generally recommended that the deformed shape be sketched first so that its correlation with other factors in the investigation may be used as a check on the work. The following examples illustrate the process of investigation for simple cantilever frames.

Example 1. Find the components of the reactions and draw the free-body diagrams, shear and moment diagrams, and the deformed shape of the frame shown in Figure 23-2a.

Solution: The first step is the determination of the reactions. Considering the free-body diagram of the whole frame (Figure 23-2b), we compute the reactions as follows:

$$\Sigma F = 0 = +8 - R_V, \qquad R_V = 8 \text{ k (up)}$$

and, with respect to the support, labeled 0

$$\Sigma M_0 = 0 = M_R - (8 \times 4), \qquad M_R = 32 \text{ k-ft (counterclockwise)}$$

Note that the sense, or sign, of the reaction components is visualized from the logical development of the free-body diagram.

Consideration of the free-body diagrams of the individual members will yield the actions required to be transmitted by the moment connection. These may be computed by application of the conditions for equilibrium for either of the members of the frame. Note that the sense of the force and moment is opposite for the two members, simply indicating that what one does to the other is the opposite of what is done to it.

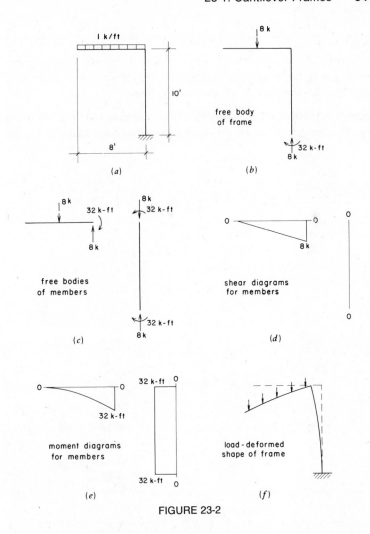

FIGURE 23-2

In this example there is no shear in the vertical member. As a result, there is no variation in the moment from the top to the bottom of the member. The free-body diagram of the member, the shear and moment diagrams, and the deformed shape should all corroborate this fact.

The shear and moment diagrams for the horizontal member are simply those for a cantilever beam. (See Figure 8-10, Case 6.)

It is possible with this example, as with many simple frames, to visualize the nature of the deformed shape without recourse to any mathematical computations. It is advisable to do so, and to continually check during the work that individual computations are logical with regard to the nature of the deformed structure.

Example 2. Find the components of the reactions and draw the shear and moment diagrams and the deformed shape of the frame in Figure 23-3a.

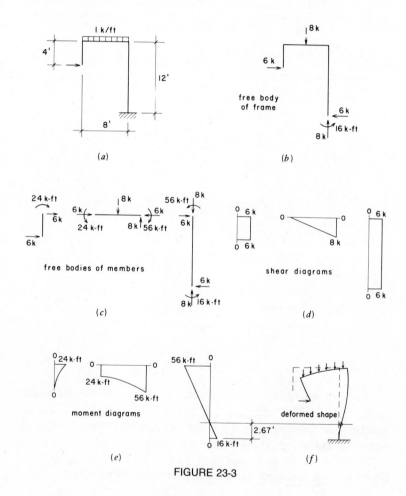

FIGURE 23-3

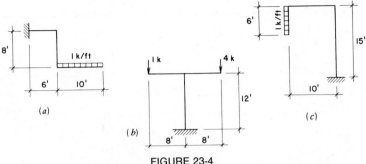

FIGURE 23-4

Solution: In this frame there are three reaction components required for stability, since the loads and reactions constitute a general coplanar force system. Using the free-body diagram of the whole frame (Figure 23-3b), the three conditions for equilibrium for a coplanar system are used to find the horizontal and vertical reaction components and the moment component. If necessary the reaction force components could be combined into a single-force vector, although this is seldom required for design purposes.

Note that the inflection occurs in the larger vertical member because the moment of the horizontal load about the support is greater than that of the vertical load. In this case, this computation must be done before the deformed shape can be accurately drawn.

The reader should verify that the free-body diagrams of the individual members are truly in equilibrium and that there is the required correlation between all the diagrams.

Problems 23-1-A-B-C. For the frames shown in Figure 23-4, find the components of the reactions, draw the free-body diagrams of the whole frame and the individual members, draw the shear and moment diagrams for the individual members, and sketch the deformed shape of the loaded structure.

23-2. Single-Bent Frames

Figure 23-5 shows two possibilities for a single-bent rigid frame. At (a) the frame has pinned bases for the columns, resulting in the form of deformation under loading as shown in Figure 23-5c, and the reaction components as shown in the free-body diagram for the whole frame in Figure 23-5e. The second frame (Figure 23-5b) has fixed bases

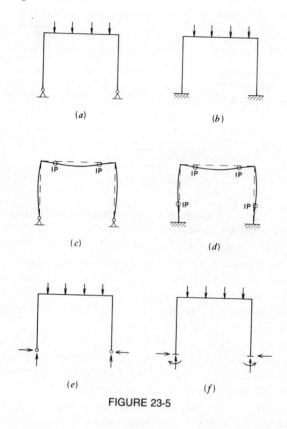

FIGURE 23-5

for the columns, resulting in the slightly modified behavior indicated. These are the two most common forms of single-bent frames, the choice of the column base condition depending on a number of design factors for each case.

Both of the frames shown in Figure 23-5 are statically indeterminate, and their investigation is beyond the scope of work in this book. The following examples of single-bent frames consist of frames with combinations of support and internal conditions that make the frames statically determinate. These conditions are technically achievable, but a bit on the weird side for practical use. We offer them here simply as exercises within the scope of our readers so that some experience in investigation may be gained.

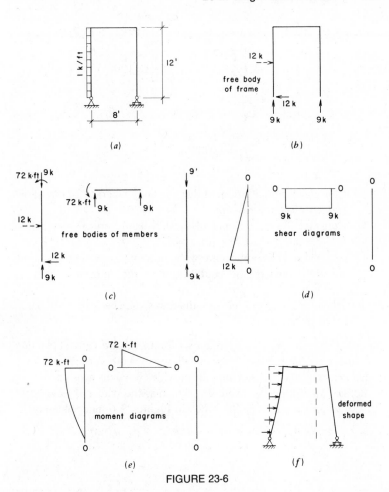

FIGURE 23-6

Example. Investigate the frame shown in Figure 23-6 for the reactions and internal conditions.

Solution: The typical elements of investigation, as illustrated for the preceding examples, are shown in the figure. The suggested procedure for the work is as follows:

1. Sketch the deflected shape. (A little tricky in this case, but a good exercise.)

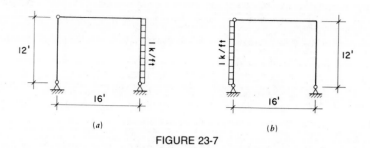

FIGURE 23-7

2. Consider the equilibrium of the free-body diagram for the whole frame to find the reactions.
3. Consider the equilibrium of the left-hand vertical member to find the internal actions at its top.
4. Proceed to the equilibrium of the horizontal member.
5. Finally, consider the equilibrium of the right-hand vertical member.
6. Draw the shear and moment diagrams and check for correlation of all the work.

Note that the right-hand support allows for an upward vertical reaction only, whereas the left-hand support allows for both vertical and horizontal components. Neither support provides moment resistance.

Before attempting the exercise problems, the reader is advised to attempt to produce the results shown in Figure 23-6 independently.

Problems 23-2A-B. Investigate the frames shown in Figure 23-7 for reactions and internal conditions, using the procedure shown for the preceding examples.

Answers to Selected Problems

||||||||||||||||||||||||||||||||||||||||||||||

The answers that follow are for those problems marked with an asterisk (*) in the text. In general, numerical answers are given only to three or four digits, and agreement in the last digit may vary with different levels of accuracy in the progress of the computations. Where structural design is involved, there may be more than one satisfactory answer to many problems; in such cases the answer given is only one of the possible answers.

Confidence in self-checking is best gained when the answer is given. For this reason we have provided the answers here. However, for a better test of developing computational skills it is best to attempt to find the answer on one's own and then to verify it with that given here.

Chapter 2

2-7-C. $R = 94.87$ lb, downward to the right, 18.4° from the horizontal

2-7-E. $R = 100$ lb, downward to the left, 53.1° from the horizontal

2-9-A. $R = 58.2$ lb, downward to the right, 7.5° from the horizontal

2-19-B. $R = 82.7$ lb, horizontally to the right

2-19-E. $R = 216$ lb, downward to the left at 86.8° from the horizontal, crossing the line at 0.14 ft to the left of the vertical force

2-19-H.   Mechanical couple

2-20-D. $R_1 = 2000$ lb, $R_2 = 6000$ lb

2-24-A. $BJ = -5$ k, $JI = +4.33$ k, $JK = -1$ k, $MN = +1.73$ k, $CK = -5$ k, $LI = +3.46$ k, $KL = +1.32$ k, $DM = -4$ k, $NI = +2.60$ k, $LM = -1.5$ k

Chapter 3

3-6-D.   At point A, 133 lb; at point B, 100 lb; at point C, 400 lb; at point D, 100 lb

Chapter 4

4-6-C. $R_1 = 2620$ lb [11.589 kN], $R_2 = 2980$ lb [13.051 kN]

4-6-E. $R_1 = 4103$ lb [18.128 kN], $R_2 = 7997$ lb [35.276 kN]

4-7-A.   A force in excess of 111.1 lb [0.5 kN]

4-7-G. $R_1 = 0$ lb, $R_2 = 1600$ lb [7.2 kN]

4-7-J.   4.117 ft [1.236 m]

Chapter 5

5-10-B.   16 ft 0.119 in. [4.8 m + 2.95 mm, or 4802.95 mm]

5-10-D.   16 ft 0.345 in. [4808.56 mm]

5-13-C.   Yes, capacity is 30.25 k [135 kN]

Chapter 6

6-1-A. $c_y = 2.6$ in. [65 mm]

6-1-D. $c_x = 1.293$ in. [32.2 mm], $c_y = 3.42$ in. [85.2 mm]

6-3-B.   205.3 in.^{4} [85.4 × 10^6 mm^4]

6-3-F.   682.3 in.^{4} [284 × 10^6 mm^4]

6-4-A.   399.8 in.^{4} [166 × 10^6 mm^4]

6-4-B.   420 in.^{4} [175 × 10^6 mm^4]

6-6-C.   5.61 in. [142 mm]

Chapter 7

7-4-D.   Maximum shear = 14 kips [61.7 kN]; zero shear at 6 ft [1.8 m] from the left end

7-4-E.   Maximum shear = 3333 lb [14.6 kN]; zero shear at 5.33 ft [1.60 m] from the right support

7-5-B.   No, actual stress = 92 psi, allowable stress = 85 psi

7-6-B.   10,938 psi [75.4 MPa]

Chapter 8

8-4-A.   Maximum moment = 48 k-ft [64.08 kN-m]

8-11-C.   Maximum shear = 7290 lb [32.4 kN]; maximum moment = 42,100 ft-lb [56.1 kN-m]; inflection point at 1.23 ft [0.37 m] to left of right support

Chapter 9

9-3-A. R_1 = 7.67 k [33.35 kN]; R_2 = 35.58 k [154.79 kN]; R_3 = 12.75 k [55.46 kN]

9-6-B.   Maximum shear = 5 k [13.5 kN]; maximum moment = 20 k-ft [27 kN-m]; inflection point is 15 ft [4.5 m] from left support

Chapter 10

10-6-B.   0.36 in. [9.1 mm]

10-6-D.   No; computed deflection = 0.296 in. [7.65 mm], limit is 0.4 in. [10 mm]

Chapter 11

11-5-A.   Yes; maximum computed stress is 22 ksi [152 MPa]

11-5-C.   (*a*) *W* = 8.8 kips [39 kN]; (*b*) *W* = 4.6 kips [20 kN]

11-6-C.   20.5 kips [93 kN]

11-7-C.   Use *W* 16 × 36

11-7-E.   Use 6 × 14

Chapter 12

12-2-A.   21.2 kips [96 kN]

Chapter 13

13-2-A.   7.82 kips [35 kN]

13-2-E.   8 × 8

13-3-E.   430 kips [1913 kN]

Chapter 14

14-6-A.   21 kips (limited by bolts) [90 kN (limited by net area)]

14-9-A. L_1 = 11 in. [265 mm]; L_2 = 5 in. [110 mm]

Chapter 15

15-2-B.   3820 psi [27.24 MPa]

Chapter 19

19-3-C.   0.23 lb [1.02 N]

19-3-D.   Will not slide or tip

Chapter 20

20-3-B. f = 883 psi [6.16 MPa]; v = 321 psi [2.24 MPa]

Chapter 21

21-1-B. T = 125 lb, C = 46.9 lb

21-3-C. P_v = 33.3 lb downward, P_h = 0, compression force in leg = 83.3 lb

Index

||||||||||||||||||||||||||||||||||||||||||||||||